life ~~leted~~

my life, ~~deleted~~

A MEMOIR

SCOTT BOLZAN, JOAN BOLZAN

AND CAITLIN ROTHER

HarperOne

An Imprint of HarperCollinsPublishers

HarperOne

HarperCollins books may be purchased for educational, business, or sales promotional use. For information please e-mail the Special Markets Department at SPsales@harpercollins.com.

HarperCollins website: http://www.harpercollins.com

HarperCollins®, ■®, and HarperOne™ are trademarks of HarperCollins Publishers

FIRST HARPERCOLLINS PAPERBACK EDITION PUBLISHED IN 2012

Library of Congress Cataloging-in-Publication Data is available upon request.
ISBN 978–0–06–202548–7

12 13 14 15 16   RRD(H)   10 9 8 7 6 5 4 3 2 1

To my loving family,

who gave me the will to move forward.

my life, ~~deleted~~

1

THEY TELL ME that the morning of December 17, 2008, started out just like any other.

I routinely arrived at my ninth-floor office in the Hayden Ferry Lakeside building at 5:00 A.M. to beat the traffic and get a head start on the day. There, fueled by several cups of coffee from the break room, I would spend a couple hours of quiet time, going through emails as I watched the morning sunlight catch the water on Tempe Town Lake and brighten the Camelback and Superstition Mountains in the distance. With the economy in free fall and corporate executives realizing it was politically incorrect to fly around in private jets like they used to, I was also working to refocus my airplane management company, Legendary Jets, in new directions.

This was not the first time I'd been forced to make an adjustment in my business life—everything from a simple but different marketing tact to a major career shift—and I was sure it wouldn't be the last. After playing professional football for several years in my twenties, I'd become a financial planner and a pilot then went on to form an aviation charter company that flew corporate executives, entertainers, and organ transplant teams around the country. In less than a year my company had risen from obscurity to being chosen to fly the heart in for the first heart transplant surgery at the Mayo Clinic Hospital in Arizona in October 2005.

I soon realized, however, that I wanted to return to my original plan to build relationships with repeat clients rather than always looking for the next new customer. So I sold the charter company in February 2008 and reorganized, retaining the jet management aspect of the business. In September, after the economy took the worst nosedive since the Great Depression, I adjusted once more by modifying my marketing approach to sell airplane trips by the hour—using "jet cards," which worked like debit cards—figuring the market would rebound and jet management would become viable once again.

The new strategy seemed to be working, and things were looking up. One client was ready to buy a block of one hundred hours, possibly later that day. We'd already verbally agreed on many of the terms, and I was putting the finishing touches on my pitch to fit his particular needs. After working away for a couple of hours that morning, I was ready to head downstairs for some designer brew and a fresh blueberry muffin from the café in the building next door, which opened at 7:00 A.M.

Joan, my wife and college sweetheart, was involved in marketing and sales for our company, but she worked mostly from home, which kept our twenty-four-year-old marriage healthy. I'd had to let my assistant, Robyn, go a couple of weeks earlier because of the recession, so our only remaining employee was our bookkeeper, Anita, who came in to the office twice a week and was due in at 9:00. That morning, with my hands full of boxes of paperwork and a bag of lemons from our tree to give her, I'd accidentally left my briefcase in the car. So as I was heading out for breakfast, I took the elevator down to the basement parking garage to retrieve my case. The quiet details of daily life.

With the long strap of my briefcase slung over my shoulder, I came back up to the first floor and was walking past the backlit neon blue glass wall when I decided to make a quick pit stop in the men's room. Most of the other people who worked in the twelve-story building didn't arrive until 8:00 or 9:00 A.M., so the entire floor was empty except for the lone security guard sitting at the opposite end of the building, near the entrance.

I pushed open the men's room door and almost immediately slipped on something greasy on the rectangular beige and gray floor tiles. Everything happened in slow motion as I felt my black leather shoes skid out from under me. As I was falling backward, my eyes ran up the beige wallpaper and cherrywood paneling to the big shiny mirror, and I saw my feet fly above my head.

I did my best to try and brace my fall behind me, but there wasn't much I could do. I don't remember hitting the floor, but my head and left shoulder took the brunt of the impact, splitting my scalp open like a ripe melon. Spanning two and a half inches across, the cut went down to the bone. Because the scalp is rich with blood vessels, the gash began to bleed profusely.

I have no idea how long I was unconscious or how many times I might have fallen again as I struggled to get up. There was nothing close for me to grab onto except the built-in metal trash receptacle, so I'm not sure how I actually managed to get to my feet, but I apparently hit my face on something in the process because I ended up with a red scrape across the bridge of my nose.

Somehow I finally managed to pull myself up from the slippery tiles and made it out the door around 7:30, where I ran into a woman heading into the ladies' room next door.

"I need help," I told her groggily, promptly retreating into the men's room.

Startled by my bloody head wound as I walked away, the woman ran around the corner and into the lobby to fetch the security guard. Appearing a few moments later, he saw me trying to stop the bleeding with a wad of paper towels, my blood mixed in with the oily substance on the floor at my feet where I'd fallen.

"What is that on the floor?" I asked him. "What did I slip on?"

Later that morning I didn't remember if or how he responded, but I did remember that he brought more paper towels to slow the bleeding until the paramedics could take over. I also remembered him talking to the janitor, who came in after him.

"You had better clean that up," he said, sending the custodian into the utility closet to get a mop.

I stumbled into the nearest stall and plunked down on one of the toilets, holding the towels against my head, until the paramedics showed up at 7:50. They laid me on a board, lifted me onto a gurney, hooked me up to an IV, and stabilized me before whisking me away, with the siren blaring. They categorized me as a Level I trauma patient, meaning I needed the most urgent level of care, and rushed me to Scottsdale Healthcare–Osborn, a hospital about eight miles and fifteen minutes away in commuter traffic.

I didn't know it yet, but I'd lost my life as I'd known it—my knowledge, my experiences, and even my identity—when my skull hit that tile floor. As I reeled with pain on the way to the hospital, I could almost feel the information draining away, leaving me in a foggy, disoriented haze. From that point on, my life was forever changed.

. . .

I was pulled out of the haze by the excruciating pain of someone feeling around with his fingers in the open wound in the back of my head. My shoulder hurt too but nowhere near as much. I was lying on a thin, stiff pad on a metal cart in the middle of a wide open room, with people milling around me, all wearing the same thing. I had no idea where I was or what was going on, only that I was sick and these people were trying to help me get better. For a big, tall man like me, it was difficult to get comfortable, especially with my feet hanging off the end of the cart. In fact, it hurt to move at all.

"You're in the emergency room," a woman said to me. "Do you remember what happened?"

"I fell," I said, stating one of the few things I could remember.

"What's your name?"

I knew what some words meant but not others, and what little I still knew was continuing to leave me. No matter how hard I tried to hold on to the information, it kept trickling away. I didn't recognize the word *name*, for example, let alone what *my* name was.

"I don't know," I said. Later I would learn that she listed my name in the chart as "Peanut Butter 77," the ER's own version of John Doe.

"Where do you hurt?" she asked.

"Here and here," I said, pointing to the back of my head, then my shoulder.

The woman squeezed points along my arms and pressed down on my stomach and chest, asking, "Does it hurt here? Or here?"

"No," I said.

"How bad is the pain on a scale from one to ten, *one* meaning it doesn't hurt too bad and *ten* being the worst pain you've ever felt?"

Building on what she said, I learned what *hurt* and *pain* meant. I still knew my numbers, and although I didn't remember what order they went in, I said "ten" because my head hurt a lot.

"You're at Scottsdale Healthcare," she said. "The doctor and I are going to take good care of you, and we're going to figure out what's going on."

I felt dizzy and sick to my stomach, and I didn't understand much of what she and the other people around me were saying. All I could do was try to piece one word and one concept together at a time and build on them, although this was very difficult because everything was so garbled. I was also having trouble retaining words I'd heard only minutes ago.

When a man pulled a bright light down into my face, burning my eyes, I figured this was the "doctor" the woman had mentioned, and, as she said, he was taking care of me. Using the same logic, I figured the women who had been taking information from me and helping me were the "nurses."

The doctor rolled me onto my side. I felt a sharp prick of pain in the back of my head, and I heard a weird muffled clicking sound as he pushed something against my skull. I later learned he had given me an injection of lidocaine and stapled my wound closed.

. . .

One thing I was sure I remembered was dropping off my wife that morning for work at Scottsdale Healthcare's outpatient surgical center, so I informed one of the nurses. She called over there and spoke to someone who happened to know Joan, a former nurse who hadn't actually worked there in more than two years. Luckily, they still had a contact number for her and were able to catch her on the way to a charity event. The nurse told her that I had fallen and hit my head, so I was "a little confused," then she walked over and handed me the wall phone next to my bed.

"It's your wife," she said.

What's a wife?

I was confused because that concept had left me by now too, so I no longer knew what a wife was or what it meant to have one. It was clear that I was supposed to talk to her, though, so I did as I was told.

"Hello?" I said noncommittally.

"Hi, honey, how're you doing?" she said.

"I—am—in—the—hospital," I said in a slow, robotic monotone. I was grasping for words, forgetting what I was trying to say as I was saying it, let alone what the question was. My head was really throbbing now, and I wanted the pain to stop. Maybe then I'd be able to think clearly.

"I'm going to be there in about twenty minutes," she said.

"Okay."

. . .

I drifted in and out while the ER staff, dressed in baggy blue pants and short-sleeved shirts, put a plastic patch around my finger and sticky patches on my chest, all of which attached to lines that hooked up to a machine.

"I'm going to take your blood pressure," one of the nurses said.

I trusted she knew what she was doing.

"We're going to give you something for the pain," another nurse said. As she put the clear liquid in the tube, I felt the pain subside a bit, but the relief didn't last long.

Next, a man called a "tech" wheeled me down the hall, explaining, "We're going to take pictures of your head and neck."

He helped move me from my bed onto a table that pulled me inside a big open machine shaped like a semicircle with a hole in the middle. Once I was inside it, he instructed me to lie still. I kept my eyes open while the table moved into the tube as it clicked and whirred.

From there they wheeled me to another room nearby, where they put the head of my bed against the wall. After getting some more pain medication, I felt a brief euphoria and calmness, which made it even harder to think or listen to the nurses. It was a fight just to keep my eyelids open because they were getting so heavy. So I gave in to the sensation and let it take me and pull me down into sleep.

. . .

I continued to drift in and out of consciousness as I lay in my corner. I didn't speak unless I was asked a direct question. I just listened and tried to absorb information, not wanting to say the wrong thing. As uncomfortable as the bed was, I was afraid to move in case I fell again. I also felt scared and confused about why everyone but me seemed to know what was going on.

Why don't I know anything, and why can't I remember anything? What happens now?

. . .

A little after 9:00 A.M., I heard a tap, tap, tapping that was getting louder when a pretty blond woman turned the corner and walked toward me. I noticed that she was dressed differently than the nurses, wearing a long-sleeved gray sweater covered with tiny pearls and sequins that caught the light, black pants, a black jacket with furry cuffs, and the shoes that made that tapping sound.

"Hi, honey," she said, which gave me a clue that I belonged to her somehow. Her tone of voice was different—more personal—than that of the nurses. From her frown and tight lips, I could see that she was upset. Still, she leaned in, gave me a hug and a kiss on the lips, enveloping me with a strange but comforting warmth. I inhaled a sweet scent, noting that she smelled much better than the other women who were taking care of me. But more important, she seemed more attentive and affectionate. I wondered who this woman was; she seemed so troubled on my behalf.

One of the other nurses who came over to talk to her at my bedside unknowingly supplied me with some answers. "You must be Jelly," she said, jokingly to explain the Peanut Butter name.

"Oh, I'm Joan, his wife," the blonde replied, trying to smile through her concern.

As I heard Joan call herself my "wife," I could tell that she too felt she was something more important to me and I to her. Hopefully in time I would piece that together as well.

The two women chatted, throwing around medical terminology that was foreign to me, but it was clear that they, at least, spoke the same language. I slowly began to pick up on some simple names of things and their meanings.

"He keeps insisting that you work for the hospital," the nurse told Joan. "In fact, he was quite adamant about it."

Joan turned toward me, looking puzzled, and said, "I've worked with you for the last two and a half years."

She knew I had a head injury, so she wasn't all that worried, but I was more puzzled than she was. I felt lost and alone. I didn't have a clue what this woman was talking about. If she'd worked for me for two years, why didn't I know her?

But Joan was clueless as well. "Well, at least he didn't forget me!" she exclaimed.

I let her think what she needed to. She would find out soon enough.

Ultimately, it took me six weeks to get up the nerve to tell her that I'd had no recollection of her as I lay in that hospital bed.

She was the woman I'd fallen in love with, married, and fathered three children with, and yet I had forgotten everything there was to know about her and our life together.

But one question was nagging me even more: Who the hell was *I*?

2

I LISTENED while the nurse asked Joan a slew of questions about my medical history, trying to determine, apparently, whether I might have gotten dizzy and fallen or if I had, in fact, slipped and fallen, as I'd told the paramedics. I listened closely, hoping to discover some telling facts about myself. This variable made a difference, I later learned, in terms of my possible diagnoses and treatment.

Although they had me on a nothing-by-mouth diet because I'd been vomiting, Joan mentioned that I'd had weight-loss surgery, with a band inserted around my stomach, so, among my other dietary restrictions, I couldn't eat much more than a cup of food at a time. I couldn't remember this, of course, but I'd already lost more than fifty pounds since I'd topped out at three hundred and seventy.

Once the nurse left, I told Joan the few snippets of what I could remember leading up to and right after the accident, then promptly forgot them.

"I remember taking one step into the bathroom and my feet flying over my head," I told her. After I fell and hit my head, I said, "I just could not get up; I kept slipping."

She was listening to me, but I still felt a need to persuade her I was telling the truth. "Look," I said, extending my palms toward her. "There's something oily on my hands."

But I could see that she was not convinced, and I could under-stand why. "Honey, there's nothing on you," she said, running her fingers over my palms. "They must have cleaned you up."

I may be confused, I thought, *but in this case I know what I am talking about.* My mind was whirling as I tried to absorb and re-tain new information, so it seemed extremely important to com-municate to Joan the few details that I could remember about my accident before they too left me. I distinctly remembered that the floor had been slippery from whatever greasy substance had been spilled there, and after I'd managed to get up, I had seen and felt it all over my hands and arms. I also remembered flashes of rubbing my hands in the ambulance, trying to determine what the slimy substance was and how to get it off me. I needed her to believe me, but my thoughts didn't settle in the most logical order. It was also frustrating to try to express myself now that there was no oil on my hands. How could I prove this to her from my hospital bed? I pulled back the blanket to see if there were oily spots on my pants and shoes, only to find that I wasn't wearing them anymore. All I had on was one of those flimsy patterned gowns that ties in the back. Where were my clothes?

I struggled to come up with some other way to prove my story. "Get my pants," I insisted. "Let me show you."

Joan reached under the bed, where she found a plastic bag that contained the black polo shirt and olive-green pants I'd been wear-ing. Once she acknowledged the oily, dark blotches, I was finally able to relax a little. Then I moved on to the other important part of the story.

"There were two men, the security guard from the front desk and a custodian in the bathroom, who were helping me and getting me paper towels to hold on my bleeding head," I said.

I told her what I'd said to the guard and relayed his direction to the custodian to clean up the mess on the floor. With all that out of the way, I was done talking for the most part. My head was killing me. The nurse kept asking how much pain I was in, and I kept say-ing "ten" because that seemed fitting. But she apparently figured it had to hurt a little bit less after the morphine she'd already given me, so when she suggested, "An eight?" I said, "Okay."

Feeling the need to protect myself from further harm or any conflict, I mostly tried to agree with what she and Joan said—anything so as not to raise more red flags than necessary over my condition. I also figured it would be best to let Joan take over and be my voice. Let her figure things out for me.

Joan was nice, but she kept asking me questions when really all I wanted to do was close my eyes and sleep.

"Are you feeling okay?" she kept asking. "Are you feeling sick?"

"Pain," was all I could manage. "It hurts."

They'd given me something for the nausea, but I was still throwing up and feeling dizzy. It also didn't help that I couldn't answer most of Joan's questions, which only made me more frustrated, embarrassed, and scared because I didn't understand *why* I didn't know the answers. I did my best to focus, to pay close attention, to listen and learn, making new connections with words and concepts whenever I could. Even so, Joan was starting to realize that my condition was worse than she'd thought.

I heard her tell the nurse and doctor that she used to work there and at another hospital as an ER nurse. That helped explain how she knew so much, such as when to put cool cloths on my head, which felt good. So did her touch.

As I nodded off, the memories of my fall and these early conversations with Joan soon faded into a blur of the emergency room chaos.

. . .

When the results of my blood work and CT scan came back normal, Dr. Douglas Smith figured I had a bad concussion and he was ready to send me home. Wherever that was.

Joan, however, seemed very uncomfortable with the idea of my being released in this state; she sensed that something else was wrong.

"He's always had a very high pain threshold," she told the doctor. "It's unusual for him to complain of so much pain."

She was also troubled, she told him, by the gaps in my memory, which didn't seem to be improving.

I was apprehensive myself. No matter what the tests said, I felt anything but normal. And I had no idea what "home" was other than it meant leaving this place where people were taking care of me and giving me medicine for my pain. I was still in too much agony to move, and I was scared of doing anything to hurt myself further.

Dr. Smith didn't seem all that concerned about my headaches, saying they were a normal symptom of a head injury like mine. They were taking steps to discharge me when around 10:00 A.M. I noticed a dark area beginning to form in the bottom of my right eye, like a black pie-shaped wedge between four and eight o'clock in my field of vision. As if the pain and memory loss weren't enough to deal with, was I now losing my sight too? I tried not to let Joan see the panic that was building inside.

"What's going on with my eye?" I said. "Part of my eye is dark. I can't see."

Joan looked even more scared than I was.

"I'll draw it for you," I said.

She grabbed the cardboard tissue box next to the bed and handed me a pen out of her purse. I turned the box upside down and drew a circle as if my field of vision was a clock with the dark hours shaded in.

She immediately motioned for a nurse, who sent the doctor over. Joan had become the interpreter of my new, small world, like a mother watching over a baby, only she wasn't aware of her role and I didn't want her to be. I watched for her reaction to determine how I was supposed to feel and to interpret the mumbo jumbo the doctors and nurses were saying. I quietly collected every piece of information I could and held them close, as if they were the bytes I needed to rebuild the master file that held my moments, knowledge, and identity—all deleted in the fall. But for now, I needed these people to keep helping me because I didn't have the faintest idea how to use those bytes to survive on my own.

I watched Joan's face as the doctor shone a penlight into my eye. She looked worried, and now the doctor seemed concerned too, which only made me feel more uneasy. Telling us he was going to call a specialist for a consultation, he left the room but returned a short time later to inform us that the doctor was busy.

"We're going to keep him and have the neuro-ophthalmologist evaluate him upstairs on the floor," Dr. Smith told Joan.

I looked at her for a translation, so she explained. "He'll be able to look in your eye and figure out what's wrong."

It seemed like forever while we waited for them to transport me to my private room. Joan kept checking with the nurses about the room status and let them know when I needed more pain medication, which was once an hour. After I complained that the morphine wasn't doing the trick, they threw in some Tylenol.

. . .

Around noon an ER nurse said my room was ready, and we waited for a tech to wheel me into the elevator and up to room 636. There I was relieved to find that my new bed was much more comfortable—larger, softer, with controls that allowed me to raise and lower the top half of my body. This was a big improvement because, while lying flat, the pain in my head was unbearable.

It was a smaller room than my corner of the ER, but it had a great box with moving pictures mounted on the wall next to the window. Joan controlled the gadget that changed the picture box for me, which I soon learned were called a remote and flat-screen TV. I gradually started to learn my previous programming likes and dislikes because she stopped when she got to one of my favorites, such as the Fox News Channel, *King of Queens,* and *Everybody Loves Raymond.*

Once I was settled in, Dr. Johnny Walker, an upbeat doctor in his midforties, came in and introduced himself as the primary care doctor who would be coordinating my treatment. "He seems to have suffered a severe concussion, but everything should start coming back to him in the next few days," he said, directing his comments to Joan and me as I watched both of their faces. "The neurologist and neuro-ophthalmologist are on their way to see him."

Walker asked me the same questions I'd heard before, and little by little I was learning some of the right answers. I didn't want to look stupid, so I listened most closely to things I knew I'd be asked again.

"What is your name?" Walker asked.

By now I knew that one, so I told him.

"Who is the president?"

"Bush," I said.

"Well, that's close. Barack Obama just won the election."

I didn't really understand what that meant or what a president was, but I mentally chalked up the correct answer. At least now I knew what to say next time.

"What is your birth date?"

"February 23, 1960," I said, repeating the same answer I'd been giving.

Out of the corner of my eye, I saw Joan mouthing to the nurse, "That's my birthday." Catching that, I internally noted that I was still wrong and that I needed to figure out the right answer, my own birthday. But at that point I didn't realize that I'd gotten her birth year wrong as well.

Walker had me do a series of tests that he and the other doctors kept repeating—pushing my hands against his, squeezing his fingers, and pushing my feet against his hands.

"For now, just focus on resting and getting rid of the pain," he said.

Right before he left for rounds, he told us he'd Googled me and noted that I'd played in the NFL. Baffled by the terms *Google* and *NFL*, I just played along.

"You're a tough guy," he told me. "You'll do fine."

I understood the gist of what he was saying and wanted to be-lieve him, but somehow I still wasn't convinced.

After he left I asked Joan about the terms he'd used. "What's the NFL?"

"The National Football League," she said. "You played profes-sional football."

That still meant nothing to me, any more than Joan's explana-tion about Google being a search site on the Internet, because I didn't know what those words meant either. So I just filed them away to figure out later.

I got some answers that evening when I came across the NFL playoffs as I was flipping through the TV channels.

"Oh, there's football," Joan said.

"I played that? Get out of here," I said, skeptical that I could have done something shown on TV.

"Did I play on television?"

"Yes," she said, which also left me amazed.

Oddly enough, I still knew certain football terms and rules such as *offside, penalty,* and *holding;* I also understood what a touch-down was and what was happening on the field. But I had no recollection of playing, how many players were on a team, or any players' names.

. . .

The neurologist, Dr. Richard Goodell, showed up about an hour after Walker left. He checked out my eyes and threw a bunch of medical terms at us that went right over my head, but luckily Joan understood them. Although he couldn't see a hemorrhage in the retina or vitreous, he said that was likely the cause of what he called my "posttraumatic visual defect." Less likely, he said, was that the fall had caused a trauma to my occipital lobe, the part of the brain that controls what you see. After he left Joan tried to explain what he said, but her explanation still left me mostly in the dark. Now that I was in a private room, the nurses hooked me up to a morphine pump, which helped bring the pain down to a slightly more manageable level. And there, on the wall, was a chart showing the pain scale—with a happy face at zero and a progression of crankier expressions from one through nine, with a crying face at ten—which helped me keep things straight. It was frustrating to need the drug, but I felt that if I could get the pain to stay at a five or six rather than sliding back up to an eight or nine between doses, my mind would clear a bit and I'd be able to understand more of what was going on around me.

Joan had been continuing to ask me questions between doctor visits and was slowly seeing just how little I knew, which prompted her to challenge Goodell's prognosis.

"But he has these large memory gaps," she said. "He doesn't know our business, where we live, or any details about our twenty-four years together."

And she didn't even know the half of it. Nonetheless, Goodell didn't alter his prognosis, saying the neuro-ophthalmologist would be by shortly. All business, he was the kind of doctor who got annoyed when Joan's cell phone rang while he was talking, and he left us with the same reassurance as the other doctors: I had a bad concussion, and my memory should come back within a couple of weeks. The headaches could last as long as two years, he said, but their intensity should lessen. I didn't understand the difference between two weeks, two years, and twenty-four years, but based on Joan's comment, it sounded like we'd known each other for a long time.

At 3:15 Dr. B. K. Suedekum, the neuro-ophthalmologist, brought in some special equipment and put drops in my eye. After examining me, he said he couldn't tell for sure either, but he thought I might have a retinal tear, so he wanted a retinal specialist to take a look at me.

Clueless once again, I looked over at Joan to figure out what that meant. Seeing her crying for the first time was a clear enough indication that the prognosis couldn't be good. I'd learned as much from the pain chart.

. . .

Given the severity of the situation, Joan thought it best to let our family and my bookkeeper know I was in the hospital. When she told me that Grant and Taylor were coming to visit, I had no idea who she was talking about.

"You didn't forget your kids, did you?" Joan asked.

I nodded reluctantly.

"Well, do you remember what they look like?"

I shook my head, so Joan picked up the rectangular metal thing she'd been talking into and pointed to a photo of Grant, then opened a flat, folded pouch for a picture of Taylor.

Joan started telling me about them, that Taylor was sixteen and a cheerleader. "Everybody says she's a mini-me," she said.

Grant was nineteen, she said, pausing. "He's more quiet. Used to ride motocross. Very competitive. He's really tall, like you."

"Are we close?" I asked.

"Yeah, we're a very close family," she said. "We deal with whatever comes our way. We handle it."

Despite my attempts to hide it, the vast extent of my memory loss was starting to sink in for Joan, so she tried to give me the information I needed in simple terms. She informed me that my children loved me and that I'd always been a good and active father. Trying to ease my anxiety, she underscored that head injuries could cause a temporary memory loss. But honestly, I was in so much pain I wasn't paying much attention. I was barely able to acknowledge what she said.

· · ·

Around 2:30 P.M. a beautiful athletic blond teenager walked into the room and started crying as soon she saw me lying there, still hooked up to the blood pressure monitor and morphine drip.

"Daddy!" she said, coming over to hug me. I hugged her back and gave her a kiss. Even though I felt no emotional attachment to her, I did feel what I can only guess was an instinctual urge to comfort and protect her. That said, I couldn't tell if she was crying because she was sad or scared or both.

A very tall—six feet three inches—and broad-shouldered young man with light brown hair came in behind her, waiting his turn to lean over and give me a somewhat cooler partial hug and pat on the back. I didn't feel the same level of affection from him as from the girl, but it was obvious from the way they touched and interacted with me that *family* meant something more to me, and vice versa, than did the doctors or nurses, who showed me a brief and much more superficial level of concern.

Taylor seemed afraid to come too close at first, not wanting to cause me any more pain.

"Let me see the back of your head," Grant said.

I turned to show him where they'd shaved off my hair to put in the four staples, which seemed to satisfy the young man in some way.

"Do you want to see?" Grant asked his sister enthusiastically. She screwed up her face to say no but couldn't resist looking anyway.

"Yeah, when I had my staples, they drove me crazy, itching," he said, as if we now shared a special bond. I noticed that Grant had both of his ears pierced with dime-size objects.

Taylor really was a mini-Joan, curling up next to me on the bed and asking me a bunch of questions, exploring how much I really didn't know.

"Do you know we have two dogs?" she asked.

I could answer that one. "Yeah," I replied. "We have a yellow and a brown."

"No, the yellow lab, Cody, died," she corrected me. "We have the black one, Aspen."

I shook my head blankly to convey that I didn't remember that. "And the brown one is Anthony," I offered.

"Dad!" Taylor said, looking to her mom and brother for support.

"What?" I asked, confused. Taylor seemed upset, but Joan and Grant were laughing.

"That's Taylor's boyfriend," Joan explained. "The brown one's name is Mocha."

It was strange how, occasionally, these few little glimpses of memory came back to me, but they were all scrambled up. Later

Joan told me that I often teased Anthony about being Hispanic but didn't mean anything by it. I also found out later that Anthony had just crashed Taylor's car the day before my accident, and I had been understandably quite angry, so they were both hoping I would forget about that too.

While we were talking, Taylor and Grant pulled out these flat contraptions that looked like the one Joan had been talking into. I watched the kids staring down and moving their thumbs around on them, and wanting to fit in, I figured I should have one too. Even if I didn't know how to use it.

"Do I have one of those?" I asked.

"Yes, right here," Joan said, pulling mine out of her purse, where she also had been keeping my watch and wedding ring, which the nurse had given to her. The screen on my gizmo, which they told me was called a BlackBerry, said I had missed two calls.

"Do those names look familiar?" Joan asked about the people who had called me.

"No," I said, wishing they did.

As I started touching the different buttons, trying to see how it worked, Taylor sat perched on the bed beside me, clicking through the various photos of people and airplanes I had stored in my phone. She kept asking if I recognized any of them, but none of them triggered a single association, emotional or otherwise.

From the way Taylor smiled at me, touched me with so much care and compassion, I could tell that we must have had a very special relationship, and I so wished that I could have felt more for her. I sensed I was supposed to feel an emotional bond with my wife and children, but I couldn't access any of the good times and tender moments I'd shared with them. My memory bin was like a giant black hole of nothing. All I had to go on was what was going on in front of me now—their warm touches and their worried expressions.

Even though my family was there beside me, I still felt very alone, as if I were trapped in a person I no longer knew. The more questions they asked and I couldn't answer, the more panicky and overwhelmed I felt. I wondered if perhaps my memory would never improve. *Is this the way it is always going to be? How could this happen to me?* And, as if I'd had a choice, I wondered how I could have let this happen to my family. Not remembering that I'd been a strong patriarchal figure, I had no sense of how much I meant to them or how much they had always relied on me for love and support. Still, deep down, I knew I wasn't the same person. I feared I would never be normal again.

Taylor had to go to cheer practice, and Grant had to drive her there, so I tried to send them off reassured by parroting back what the doctors had been telling me: I was going to be okay, they shouldn't worry, I'd be coming home soon. By this point, I'd picked up that *home* was someplace they were going once they left the hospital.

As strange and unimaginable as it was to forget my children and the love of my life, I was trying to stay positive and have faith in the doctors' prediction that my memory would return. I clung to that hope, but after losing part of my sight, I was also scared that my condition might get even worse.

. . .

Dr. Derek Kunimoto, the retinal specialist, stopped in about 6:00. He was young, articulate, friendly, and confident, and he seemed very capable, which helped me feel more confident too. After shining a light in my eye with a silver tube and adjusting a little wheel, he gave us the first piece of good news we'd heard so far: my retina was still intact. He said he didn't know what had caused my dark spot but suggested that I might have a microhemorrhage in my optic nerve, which Joan later explained meant bleeding in my eye that was too small for us to see.

After I was discharged, he said, I should go to another specialist for additional testing and an opinion on whether my full vision would ultimately return.

Joan looked relieved after he left. "Thank God," she said. It was nice to finally see her happy.

I liked having her there. I had to trust someone, and she was by far the best candidate. But now that the kids had left, saying they were going "home," I figured she would want to leave too, so I thought I should repeat what the kids said and send her off. Even though I really didn't want her to go.

"Go home," I said.

Joan could see from my lost expression that I was just trying to please her. "No," she said. "As long as you're in the hospital, I'm not going to leave this room."

She told the staff she was spending the night, so they brought in a foldout chair that flattened into a bed, which they positioned in the small space between my bed and the wall, which had a tiny bathroom on the other side of it.

I still didn't understand who she was to me and why she cared so much for me, but if I felt alone with her there, I didn't even want to imagine what it would feel like with her gone.

. . .

I got little sleep that night because the nurses came in every two to four hours to shine lights into my eyes and check my pupils, ask for my name and birth date, and have me do the push-pull tests. Every time one of them woke me, Joan got up too, asking them more questions, fluffing my pillow, and putting cool wash-cloths on my forehead.

I was still throwing up, so Joan asked the nurse to try switching my nausea medications. She did this assertively but without anger-ing them, and the new meds finally quieted my stomach. It was reassuring to have someone so medically informed to advocate for my needs.

As morning approached Joan ordered breakfast for both of us—scrambled eggs and oatmeal for me, something my stomach would accept.

"That should be easy for you to swallow," she said.

Before breakfast we tried to formulate the questions we were going to ask Dr. Walker. Joan was growing increasingly concerned about the severity of my incessant pain and my profound memory loss, which only served to heighten my own anxiety.

. . .

When Dr. Walker asked how I was doing, I told him my headaches were still unbearable. He asked me the same list of questions, and I was still giving him Joan's birthday.

"I know that's not it, but that's the only one I know," I told him.

When he asked me who the president was, I thought I knew the answer this time.

"Barack," I said.

"That's his first name. It's Barack Obama," he said. "You're getting closer."

He also asked me for the date and day of the week and the name of the hospital. I was still getting those answers wrong, but after he left I noticed that the nurses had written the date and day of the week on the whiteboard across from my bed. So when Dr. Goodell asked me those questions later that morning, I looked at the board and was able to give him the correct answers. I don't think I was really fooling anyone, but at the time I thought I was outsmarting everyone.

After Joan reiterated her concerns about the memory gaps, Goodell tried to reassure her that my confusion would likely resolve in a few days to a week. However, he finally agreed to order an MRI to see if they could find something that wouldn't show up on the CT scan.

. . .

Not wanting to leave me alone, Joan stayed while we waited for the test, describing past events and sharing more of our family history. Still on the morphine drip, I listened as best I could, but I kept fading in and out as she talked, so she waited patiently until I came to before starting up again.

She told me, for example, that when we met at Northern Illinois University, she was on the gymnastics team and I was on the football team. I'd gone on to play professional football in the NFL for the New England Patriots and Cleveland Browns. None of this meant anything to me; I just nodded and tried to take it all in.

The television was always on, so we often took breaks from the conversation to watch the shows I used to like. As she was flipping around on the remote, she found a Blue Angels flyover on one of the education channels. I was intrigued by the way the planes flew so close together in a V formation, spinning around without crashing into each other.

"That would be a cool job, to be a pilot," I said.

"You are one," Joan said.

"I am?" I asked. "I don't remember that." Tears welled up in my eyes as the magnitude of this hit me. I not only didn't recall the little things, I couldn't even remember a high-powered activity that required experience, skill, and a love I'd probably felt my entire life.

"Not that kind of pilot," Joan said quickly. Not wanting to overwhelm me with information, she briefly explained that we'd worked together for the past several years on two aviation businesses—one that chartered jets, which we'd recently sold, and one that managed and maintained planes for other companies.

By then Joan was crying too. I didn't know exactly why but thought maybe we were thinking the same thing—how could I have forgotten so much just from hitting my head? It wasn't until about a month later that I learned that if my vision problem didn't clear up, I would never fly again. That's what Joan was thinking at that moment; she just didn't have the heart to tell me.

Joan stood by my bed, kissed my forehead, and rubbed my chest through the gown. I didn't really like being touched like that, but I went along with it.

"I know how I can make you feel better," she said, moving her hand down under the hospital covers and playfully touching my private area. Startled, I batted her hand away, wondering what the hell she was doing. I felt uncomfortable, that her hand didn't belong there. Joan looked surprised, as if this was completely out of character for me, but I was in too much pain to worry about her feelings. I had unwittingly become a forty-six-year-old virgin who didn't even know what sex was.

. . .

When Grant showed up around lunchtime, Joan went home to shower and change. She told me she was dressed up because she'd been heading to a fancy charity luncheon—whatever that was— when she got the call to come to the ER.

Grant described more memories, trying to find a trigger to retrieve some of the moments we'd shared while he was growing up. He told me that he'd started riding motocross dirt bikes at twelve and that I used to watch him compete in hockey and motocross. He competed in several national races and at the pro level in Arizona, quitting when he was eighteen, just a year or so ago.

"Do you remember any of that?" he asked hopefully.

"No, I'm sorry," I said.

It was clear from his disappointed expression that this wasn't what he wanted to hear, but I was trying to be honest with him. I was starting to pick up on how much my memory loss was making my family sad; it was as if I'd taken something precious away from them by forgetting the positive events that had shaped our relationships and strengthened the bonds between us. And no matter what the doctors said, none of us knew if we would ever get that back.

Valiantly, my son asked more questions, still searching for something I could remember—anything—until, drugged and fatigued, I dozed off.

When I came to, Grant was curled up on the foldout where Joan had slept the night before, sobbing. I didn't understand why a young man would be so emotional. It seemed a bit over the top; it wasn't like I was going to die from this head injury.

"Why are you crying?" I asked.

Grant sniffed and grabbed a tissue to blow his nose. "It just makes me sad that you don't remember anything that we did together," he said.

I didn't know what else to say other than to repeat the doctors' optimistic prognosis. "It'll get better," I offered.

This seemed to calm him down, and I felt I'd done all I could. So, like any two typical men, we stopped talking about our feelings and watched TV in silence.

. . .

Joan returned soon afterward, and Taylor showed up after school later that afternoon.

Joan, who kept leaving the room to make phone calls, told me she still hadn't heard from Thomas. I didn't know she was talking about my business colleague, but I eventually got the picture: she, Taylor, and I were all supposed to take Thomas's private plane to Las Vegas in a couple of days to watch Taylor's cheerleading team perform at a national championship. I'd seen cheerleaders on TV during the playoffs, but Joan said Taylor's team did moves that were more like dancing and gymnastics—two more terms for me to tuck away and figure out later.

After months of training, Taylor was torn between wanting to go and staying home until I got out of the hospital. She didn't want to go without us. As she tried to describe her conflicted feelings, she broke into tears. "I don't want to leave Dad," she said. "I'm scared. I don't like that he doesn't know anything."

She was worried, she said, because I wasn't bouncing back like I usually did. Joan told me that I'd had nine surgeries on my ankles, knees, and shoulders, and I'd usually felt well enough to stop at the office on my way home from the hospital. I honestly didn't know what to think about the man I used to be because everything I knew about him came from stories like these, filtered through my family's perceptions. That said, they were all I had to go on. My new life depended on them.

Joan took Taylor into the hall, but I could still hear them talking. "This is your national competition," she said. "Your team is counting on you. He's going to be discharged. He's going to be fine, and we're just going to go home."

I would soon learn that Taylor had been on the team for more than eight years, she'd been practicing several days a week for this contest, and she was one of the best on her team. Joan had gone to most of Taylor's competitions with her, but the three of us usually went to this national event together to cheer her on.

Joan continued to juggle calls with the charter company handling the flight and also with the other family that was supposed to fly with us as she developed a contingency plan. She kept me abreast of what was going on, but I maintained my poker face, not revealing that I didn't know any of the people she was talking to. Oddly enough, I still had my critical thinking skills and a vague sense of how some things worked, so I was able to suggest other options for Taylor, such as taking a later flight after I was released. But often when Joan thought I was exercising my previous problem-solving skills, I was actually just parroting back what I'd just heard her—or someone on TV—saying.

For example, when I reassured Taylor that it was okay for her to go on her trip, I was actually reinforcing the parental message I'd heard Joan delivering to her in the hallway. "You should go be with your team," I said. "Make us all proud, and don't worry about me. I'm going to be just fine."

When I told her to focus on doing well for the team and to keep her mind off the stresses that my injury was causing her, I later wondered if I'd been somehow drawing on my years of team sports and leadership as a captain even though I couldn't remember a single play on the field. Taylor, not entirely convinced, went to the gym to practice.

. . .

Around 7:00 P.M. there was a knock on the door.

"There he is," a big booming voice said. "Scottie, what are you doing here?"

The voice came from a fit-looking black man in his midfifties. His graying, closely cropped hair was balding in spots, he was dressed in casual business attire, and both he and the heavyset black woman who came in behind him looked concerned.

Feeling the hair stand up on the back of my neck, I sat right up in bed. I didn't understand why this guy was coming into my room unannounced, and I didn't like feeling unprepared for this visitor.

How do these people know me?

With nothing to rely on but my friend-or-foe senses, I felt that this guy was the latter, and I wanted him out as soon as possible.

"JD," Joan said, "I don't think he knows you."

"Oh, he knows me," the man said, shrugging off Joan's remark.

I didn't like the way he ignored Joan's attempt to smooth things over, which put me off even more. I didn't like his attitude, and I felt my fear turning to anger.

The couple stayed for about fifteen minutes, during which Joan explained what had happened to me.

JD had a big personality and wasn't the type of guy you'd forget. Still, I had no recollection of him. Nonetheless, as he was talking, I tried to act as if nothing was wrong, and when he asked me to pray with him, I didn't refuse. But as he closed his eyes, bowed his head, and started to pray to the "Heavenly Father," I left my eyes open and kept a close watch over him.

After he and his wife left, I felt relieved.

"Who was that?" I asked Joan.

How many more people am I going to see that I don't know? This could be endless.

Joan explained that he was a recent business acquaintance who had been a wide receiver in the NFL and was now a minister; this woman was his wife. We hadn't thought to ask how he'd heard about my accident, but we assumed that he must have called my office. We'd had a recent business disagreement, she said, and she was as surprised to see him as I was. It was curious that I seemed to have retained my emotional memory of him and nothing else. Other than my medical issues, this was the most anxiety-provoking episode I'd had in the hospital so far.

. . .

Finally at 2:30 A.M., on our third day in the hospital, the attendants came to take me downstairs for my MRI. Because this was a trauma center, the machine was in constant use, and this was the first available slot. It was reassuring that Joan had held to her promise to remain by my side, especially when I had to have tests in the middle of the night.

I was fine in the elevator, but when they tried to put me into that small narrow tunnel, I became irritated, fearful, and combative.

"There is no way you are putting me in that tube," I said.

Concerned that the outburst could further damage my brain, one of the staff went to fetch Joan to see if she could get me to agree to the test. But I was having so many problems expressing myself that my fear had morphed into anger by the time she arrived. I was so furious that my hands were turning white as I gripped the sides of the metal cart. I sat upright, thinking that if Joan tried to make me go inside that machine she wasn't my friend after all.

"There has got to be a different machine that they can use because I am not going in this one!" I yelled.

After some back-and-forth—heated and adamant on my side, cajoling on theirs—it was decided to reschedule the MRI so an anesthesiologist could administer a sedative. Joan acknowledged that I'd never liked confinement, and in years past I'd been tested in an "open" MRI because my shoulders were too wide for a regular testing cylinder like this one. She seemed surprised by my extreme resistance to this important test, but she proved she was, in fact, my friend by standing by me and persuading them to listen to my concerns. About 7:00 A.M. they took me back down for the second attempt, and this time an anesthesiologist gave me a Fentanyl-Versed cocktail, which produced a sense of euphoria and relaxation for a few seconds before I fell asleep. Apparently, my shoulders *did* fit into that narrow tube, where they kept me for about twenty minutes, out cold.

When I came to, I was being wheeled back into my room, where Joan was waiting for me.

. . .

Within a couple of hours Dr. Walker came by to tell us that the MRI results were normal, so they were sending me home and I should follow up with Dr. Goodell. Joan and I asked a lot of the same questions about why my pain and memory weren't improving, but the answers and the prognosis were still the same: I should get my memory back within the next couple of weeks.

Then the waiting began while my discharge papers were being prepared. As the hours went by, Joan grew increasingly agitated, which only made me more uneasy and anxious. "What is taking so long?" she kept saying. "I can't believe the staff hasn't taken care of this yet."

We waited so long that we finally ordered lunch, and around 3:30 P.M. it was time to go.

"Are you ready to go home?" Joan asked.

I was scared to find out what "home" was like, but I nodded and tried to prepare myself to find out.

3

MY SHORT-TERM MEMORY seemed to be okay, so I knew it was Friday, December 19, 2008, and I was being released from the hospital with my wife, Joan, by my side. But I was racked with fear of the unknown as I tried to prepare to leave the shelter of my sixth-floor room, the only place that felt familiar. I was still having trouble finding the words to say what I wanted, and my speech was still coming out in slow, robotic monotones.

Joan had brought me a comfortable T-shirt and jeans to wear home, but like Peter Sellers's character in *Being There,* I had no idea what to expect the real world to be like because I'd only seen it on TV.

As the tech pushed me down the hallway in a wheelchair, she used a different elevator than I was used to, and I started noticing things from a different perspective. Every other time I'd gone down that hallway, I'd been lying on a gurney. Now that I was sitting upright, a cacophony of new perceptions seemed to be hurtling toward me at warp speed as the adrenaline shot through my veins. I felt dazed and beleaguered by it all.

Outside, I squinted as the tech wheeled me down the sidewalk that ran along the driveway of brick pavers shaped like miniature stop signs, which, for some reason, I still recognized. I was confused as my eyes tried to adjust to the bright Arizona sunlight. Since the accident, when almost every point of reference I'd acquired in a lifetime had been erased, I'd been completely ensconced in the hospital world, glimpsing "outside" by watching the news. Given that it was December, I had seen lots of people dealing with cold weather and snow in other parts of the country—nothing like the sunny sky above me now—but couldn't grasp the geographical differences. So I didn't know what to think when I saw people dressed in light clothing walking around on the street. I could only assume that I was supposed to get out of the wheelchair and join them for parts unknown. Alone.

But before I could ask the young tech, "Where do I go now?" Joan said something reassuring.

"I'm going to get the car. I'll be right back."

I was still unsure of her role in my life. I hoped she really was going to come back and get me, that she wasn't going to abandon me there. But I also believed that after she came back with the car, her "job" with me would end. The relief was tremendous when she pulled up in a black Chevy Tahoe, opened the passenger-side door, and invited me into the car.

. . .

I proceeded to climb into the vehicle, which seemed to be very big for Joan, who was all of five feet one inch and about one hundred and twenty-five pounds.

"Is this the car you drive?" I asked.

"No, this is Taylor's car. I drive a smaller car, a Porsche Boxster that you bought me in Carlsbad," she said.

As she went through the litany of vehicles that we owned and who in our family drove which one, I felt my brain shutting down. I couldn't process all that information. Not when I was trying to get used to being outside in the real world for the first time.

The hospital was only a twenty-minute drive from our house, but time seemed to fly by as I stared at the buildings, cars, and people we were passing, hoping that something or someone would look familiar. It was no surprise by this point, however, that nothing did. Even though I felt like we were racing along, all the other cars sped past us.

Joan told me the names of the streets and the businesses along the route, reciting the numbers of the major roads and highways nearby, but she might as well have been speaking Chinese. I had no idea what direction we were heading or what town we were in until I saw the signs saying we were entering Gilbert, where she said we lived.

"Does this look familiar?" she kept asking.

"No," I repeated.

We'd sat at many stoplights and made numerous turns when Joan finally said, "We're turning onto our street. Do you know which house is ours?"

"I have no idea," I replied.

She slowly pulled up to our one-story beige stucco house, entered the driveway, and hit a button in the car.

Well, this has to be it if the door opened.

My head throbbed with the pain and the stress of entering yet another foreign place that I was supposed to recognize.

Joan hit another button to close the garage door, and as we walked into house, she said, "Welcome home, honey."

Taylor was in the kitchen to greet us and gave me a big hug. As we walked into the living room, with its high ceilings, wide, deep armchairs, never-ending couch, and grand fireplace, all I could think was, *This is huge.*

But the enormity of it, coupled with my pounding head, was too much to bear. All I wanted to do was rest for a while and close my eyes.

"Where is the place that I can lie down in?" I asked.

"The bedroom?" Joan asked, looking at Taylor in shock.

"Yeah. How do I get there?"

Taylor grabbed my hand and proceeded to lead me back into the master suite, which was yet another huge room.

Apparently I'm a big man with big tastes.

As Joan later explained, that assumption proved to be true: I hated to feel cramped by low ceilings and small rooms. Even so, the king-size bed seemed rather expansive, standing tall in its massive bed frame, with a nightstand on either side of it. Across from the bed was a table with some books on it and a broad wooden cabinet that encased a flat-screen television. A row of windows spanned the entire north side of the room, providing a view of a stone fountain and swimming pool outside.

I changed into some sweats while Taylor and Joan pulled back the covers and helped me get into bed. Then Joan left to fill the Percocet prescription—a mix of oxycodone and Tylenol—they'd given me before leaving the hospital.

Joan had explained that we'd lived in this house for more than three years, but as I lay in that bed, where I had slept for at least a thousand nights, it felt like the very first time.

. . .

As tired as I was, I couldn't actually fall asleep, so I lay there, thinking and examining the contents of the room: the family photos hanging on the wall and the strange but comfortable-looking armchair that had cup holders and a power cord plugged into the wall. I later learned that this was a massage chair I enjoyed using to relax my back muscles.

I was puzzled by the round, flat pillow on the floor, covered with a blanket, which turned out to be the bed for our dog, the brown Lab named Mocha.

The bedroom door was closed and I could hear voices in the next room, so I decided to take advantage of being alone to look around. I padded into the master bathroom, which also seemed huge compared with the tiny toilet stall in my hospital room. It was also much nicer, with its shiny granite countertop, two sinks, and the beautiful tile on the floor, which I later learned was called travertine. The walk-in shower was shaped like a clamshell and had two shower heads—one high up and the other set low and adjustable. Now I was no genius, but even with a head injury I was able to figure out that the tall one was my side and the lower one was Joan's.

At the far end was a wall of two mirrored doors, so I opened them to find a walk-in closet the size of a small bedroom, packed with shelves and rungs of hangers spanning sixteen feet to the ceiling with shoes and garments, some hung in cellophane bags according to season. It was no mystery that Joan loved fashion, because 90 percent of the belongings in that closet were hers, with mine tucked away in one microcorner. My mind was boggled by the discovery that she owned at least one hundred pairs of shoes, each packed in its own plastic labeled container, and that the price tags were still attached to probably 20 percent of her clothing.

Why would she continue to buy shirts, sweaters, coats, pants, and skirts when she already owns more than she can wear?

This gave me a clear indication of one foundational difference between my wife and me before my accident and perhaps now even more so. But before I could jump to conclusions, I soon realized that I'd had my favorite things as well. On my side of the bathroom counter was a stack of boxes, each of which contained a shiny fancy men's watch, while Joan had only one watch box on her side. I noticed that many of mine had different-colored faces, with different types of straps, and some had lots of little knobs and buttons.

Why would anybody need all these watches when you only have one wrist to wear them on? I am wearing only one ring now, and so did Joan, but I saw men on TV wearing lots of rings, so I maybe wore more than one watch at a time?

But, as I stood there, I couldn't understand why I would need a single watch, let alone thirteen of them, when my cell phone showed me the time.

These boxes, I eventually learned, were winders for my Rolex, Baume & Mercier, Breitling, IWC, Omega, and Chase-Durer timepieces, all of which I loved and was proud to have been able to afford. But at that point I had no idea that most of them were worth thousands of dollars each or that I decided which one to wear on a given day based on what else I was wearing or where I was going.

Going through the drawers and lower cabinets, I started noticing a pattern of excess: packages with thirty rolls of toilet paper,
packs of ten toothbrushes, and half a dozen containers of the same
brand of men's deodorant. I would later find out that this was
called "buying in bulk," which was cheaper and more convenient
for people like me who hated shopping, but at the time I thought it
was way too much stuff for the two of us.

Searching further, I was surprised to find about twenty bottles
of cologne in the mirrored cabinet next to my sink.

*Are these all for me? Why would I ever need twenty when I could
wear the same one every day?*

I'd seen plenty of cologne and perfume ads on TV and in Joan's
magazines in the hospital, so I opened one of the bottles and
smelled it, pumped the spray onto my hand, and smelled it again.
It smelled good, so I grabbed another and another, spraying each of
them in the same place and smelling it again. I went over to Joan's
side and saw that she had about half as many bottles of perfume,
but that still seemed like too many for one person. I didn't test any
of hers, though, because I didn't know if I was allowed to.

Based on what I'd observed in those few minutes, I already felt
I'd been a complicated man whom I now didn't understand, and
it was going to take a lot of investigation and thought to figure
him—or me—out. Joan seemed complicated too but in a different
way, so the concept of us living together was going to be that much
harder for me to figure out.

Leaving the master suite, I went to find Joan and Taylor in the
family room and asked if they would show me the rest of the house.
I wanted to hear more about my life.

They led me past the front door to a guest bathroom on the
other side of the house and into a second family room, where, I was
told, Grant and Taylor usually hung out with their friends to watch
the large television. It was a smaller version of the TV we had in the
living room, which was much easier to watch at a distance than the
one in my hospital room.

After that we entered Taylor's bedroom, which had beautiful hardwood floors. Her bed-frame posts were hung with about thirty purses, and her shelf was covered with shiny gold trophies topped with little figures dancing and cheering. I had picked up the word "trophy" while watching the playoffs, and seeing that these said 1st Place, I figured these were Taylor's awards for cheerleading.

Next came the guest bedroom, which Joan said used to be Grant's and was the one he used when he stayed over although he, apparently, used it more than any guests. She said our nineteen-year-old had his own apartment a few miles away and worked two jobs, repairing motorcycles and delivering pizzas.

Finally we came to one last room, which contained a bed next to a U-shaped desk that was stacked with four or five computers with their screens turned on, a fax machine, and a phone. A *Ranking Arizona* magazine cover, displayed on a stand, featured a photo of Joan and some other people with text that said our company, West Jet Aircraft, had been ranked number one in 2008. On the wall was a photo of a surgical team standing next to one of our jets. "This is your home office for when you don't feel like going to your Tempe office to work," Joan explained.

I soon discovered that closets throughout the house were filled with file cabinets, office supplies, and boxes of aviation company documents. By the looks of it, I'd been a very busy and successful businessman.

Also hanging on the wall was a red and white football jersey and two newspaper articles with photos from the California Bowl in 1983. When I asked Joan about them, she said I'd worn the jersey during that game, and she'd gotten it framed for my birthday years ago.

That game must have been very important to me.

I wasn't interested in the details right then, but I later discovered that game had solidified my chances for an NFL draft in 1984 because I'd dominated the defensive tackle on the opposing team, helping us beat Cal State Fullerton, 20–13.

We passed through the living room again to head into the back-yard, where I admired the landscaping, including the L-shaped pool with its own waterfall, and the three-tiered waterfall, fire pit, lounge chairs and table in the secluded grotto at the other end. Even in December it was sunny and a pleasant sixty-five degrees out there, where, Joan explained, we'd spent many an evening with cocktails while I smoked a fine cigar.

"Wow, this is really nice," I said.

I truly was blown away—overwhelmed, really—by our home. But even after taking the full tour of the house and garden, which had been paid for with my hard work, I still felt like a guest in it. As uncomfortable as that felt, though, I knew that I'd better get used to the idea of living there with these people who called themselves my family because they were all I had. The uncertainty I felt was pretty intense, so it was good that I was heavily medicated because that helped me relax a bit. And because the meds made me so drowsy, I was forced to sit still and take frequent naps in those early days, which better enabled me to ease into my new surroundings and absorb new information in short, finite bursts.

I knew in my gut that I had to start making myself comfortable around these people so they would feel like I was getting better. I didn't know how long they were going to put up with taking care of me in my helpless state, nor did I have any concept of time to understand the duration of a marriage or a relationship. I could see from the wedding photo in our bedroom—two young kids dressed in all white and our faces stretched with sappy smiles—that I had aged quite a bit since then, so I figured they'd keep me around for a while anyway.

The last thing I wanted was for them to leave me completely on my own. I felt they would be not only crucial to retrieving my identity, but key to my very survival.

. . .

A couple of hours later Joan suggested I call my parents, whom she'd been keeping abreast of my progress in the hospital. Because the doctors said I only had a bad concussion and my memory should return within a couple of weeks, my parents saw no need to rush to my bedside. Joan told me they lived in Chicago, and I had no idea how far away that was, but they weren't a priority for me at that point. It was all I could do to deal with the people right in front of me—Joan, Taylor, and sometimes Grant—but I sensed they were important, so my plan was to put them at ease and get off the phone as quickly as possible. My head was killing me.

Wanting a point of reference, I went into my office and picked up an eight-by-ten-inch portrait of them. The tan frame was embossed with the word "Grandparents," and although they were strangers to me, at least I could put a voice to each of their faces. They looked very happy, smiling and leaning in toward each other at a restaurant table in front of a fireplace. I could see some wrapped gifts, which Joan said were to celebrate my father's seventieth birthday.

"How are you feeling?" my mother asked.

As we chatted on the phone, I stared at the dark-haired woman wearing a purple and black pantsuit and the gray-haired man in his beige sport coat and burgundy shirt, but I felt no emotional tug whatsoever.

I didn't want to alarm her, so I kept it simple, not mentioning that I had no recollection of them. "I had an accident," I said. "My head hurts."

I asked them to help Joan if she needed anything because, obviously, I couldn't. The call didn't last more than five minutes, but it felt like five hours. By the time I hung up, I was completely exhausted.

When Joan said my sister Candi was on the phone a little while later, I didn't have the energy to go through that again. I tried to wave Joan away, but she covered the mouthpiece and gave me a quick lesson in manners before placing the receiver in my hand.

"Hon, she just wants to hear from you that you're okay," she said.

I reluctantly agreed to talk to Candi, but if it had been up to me, I wouldn't have spoken to anyone else. Afterward, I figured that if

Grant had a sister and Taylor had a brother, I might have more of each and needed to prepare myself to speak to them too.

"Do I have more sisters or brothers calling?" I asked Joan.

Joan explained that I was the youngest of three children and that I had another sister, Bonnie, but no brothers. That meant I had one more call to go, which was both a pain and a relief that this one would be the last.

I 'D BEEN HOME from the hospital for two days when Joan and I debated whether to cancel our traditional extended family Christmas dinner. Typically we'd always had everyone over to our house because our long dining table could accommodate all eleven of us, including Joan's parents, my niece Jamie, her husband, Kevin, and my two young, rambunctious nephews, Noah and Aden.

"Maybe it's best we not do a big family get-together," Joan said. "Maybe it should just be us this year."

In spite of my incessant headaches, I still felt the need to see and do the same activities as everyone else in the world, partly so I could understand what "normal" was, but also to observe and participate in an important holiday with the family I was struggling to get to know.

"Don't change things just because of me," I said.

Joan said she was concerned that I wouldn't be able to take the boisterous running around that my two- and six-year-old nephews were apt to do, but I told her I could handle it.

"How much more can my head hurt?" I joked.

So Joan gave in. "Well, if you need to go lie down, you go lie down," she said.

I was curious about the meaning of Christmas and why we celebrated it, and while I was learning at a relatively rapid rate, I could take in only so much information at once, so Joan parsed it out little by little. I'd been picking up quite a bit by watching television, such as the Christmas episode of *The King of Queens,* in which Doug and Carrie Heffernan's working-class friends and family come over with gifts and everyone laughs together while eating a special dinner.

"Is this how our Christmas is?" I asked.

The holidays seemed to be a generally happy time for these and most other TV characters, and I wondered if I could even pretend to be happy considering all my pain and fear, not to mention the ongoing pangs of emptiness and loneliness.

"There will be funny moments," she said, explaining that her father, Harvey, was similar to Arthur Spooner, the father on the show played by Jerry Stiller. Like the crotchety older man with a big heart and a biting wit who lives in the Heffernans' basement, she said, Harvey tries to be funny and often blurts out whatever is on his mind, and sometimes it's inappropriate. You never know what he's going to say next.

I'd also seen the Christmas episode of *The Sopranos,* which provided quite a contrast, with their lavish spread of food and the expensive cars and jewelry they bought each other. Every time I acquired a new reference point like this, I'd question Joan or Taylor about what I could expect and where we fit in.

. . .

As the days went by, I saw Joan bringing home more and more bags—some with handles from nice department stores and some white plastic ones with a big red circle on the side from a discount store called Target. I watched her secretly pull out jackets covered with rhinestones, long necklaces, and a leather tote bag for Taylor, a number of gift cards for Grant, and packages of action figures for my nephews. I was mesmerized as she wrapped each gift with colorful paper and a shiny bow then stuck a label on it, saying who it was for and from. I was puzzled why she wrote "Santa" on some of Grant's and Taylor's gifts, so she explained all of that. I also didn't understand the purpose of this giving of gifts or why there were so many of them.

Joan said we also exchanged presents with my parents in Chicago and that she'd already sent my mother several restaurant gift cards because she and my father loved to go out to dinner. Joan had also sent her a toy Santa with a saxophone, one in a long line of goofy presents over the years. But she noted that she'd had to scale back on the home front this year because she simply didn't have the time to do her usual shopping, so there wouldn't be the same gag gifts, gold chocolate coins, or ornaments she traditionally bought for the kids. Somehow, she'd still found time to do her usual wrapping job, hiding one small item in at least five larger individually wrapped boxes, so the kids would keep pulling off the paper until they reached the payoff.

Joan clearly enjoyed this holiday, and although she kept trying to convince me that this Christmas wouldn't be any different, I knew better. I figured it would be a memorable event for us; I just wasn't sure how.

. . .

No one expected me to help with preparations for the big day, but I wanted to, thinking there was no better way for me to learn about memories I shared with my family than to relive them as best as my body would allow. So while Joan ironed the tablecloth, I helped hold it and guide it. And when I couldn't figure out how to help with a given task, I retreated to my chair and nursed my aching head.

Joan took breaks from her shopping, cleaning, and other responsibilities to spend some quality time with me in my big chair in front of the television in the living room, rubbing my head and snuggling in my lap. I appreciated the attention, but I was somewhat preoccupied with hiding the pain of my headaches and the day-to-day battle of simply trying to exist.

On December 23 she went out to run some errands, leaving Taylor and me at home to watch TV and talk about holiday rituals in the Bolzan household.

"Is there anything you and I do together?" I asked.

"We make Christmas cookies," she replied.

"Well, then, let's make Christmas cookies," I said, reminding her that she'd have to teach me what to do.

Taylor found a sugar cookie recipe on the Internet, but not wanting to leave me alone, she waited for Joan to return before she ran to the store to pick up the necessary ingredients. Joan cleaned off the center island and laid out a collection of cookie cutters— trees, angels, snowmen, and stars—then left us alone to share some daddy-and-daughter bonding time.

Taylor exhibited great patience as she walked me into the pantry to show me where we kept the flour, sugar, and vanilla. After I retrieved the bucket of bottled red and green sprinkles and tubes of frosting from a cabinet she couldn't reach, she dotingly pointed out how the lines on the measuring cups corresponded to the amounts in the recipe, demonstrating how to carefully transport the ingredients from the cups and spoons to the bowl.

"Are you sure that's right?" I kept asking. "How much was it again?"

We weren't all that careful, however, because the floor was soon covered with flour and sugar. The dogs were only too pleased to help with the cleanup.

I felt a great joy, a sense of peace and a feeling of safety with Taylor. I barely knew this beautiful, intelligent girl, and yet she was displaying a love that had clearly grown over a lifetime. For those two hours we spent making several trays of cookies, some of which we ate warm, right out of the oven, I felt as if no one else mattered to her as much as I did. She also seemed to enjoy the experience of letting me learn about her all over again—her friends, her cheer-leading, her boyfriend, and the time she split between her regular academic classes at Mesquite High School and her fashion design and merchandising courses at East Valley Institute of Technology, splitting her day between the two schools.

At one point I started to cry because I couldn't believe I didn't recall a single thing she was telling me. The realization that I had lost all those precious moments of watching her evolve into this sweet young lady made me so very sad.

"Don't worry about it," she said, hugging me. "It's only been a couple of days, Dad."

It was almost as if we'd switched roles of parent and child, a pattern that continued as she reassured me that my obviously poor job of decorating the tree-shaped cookies with frosting was better than it really was. I wasn't sure how good my artistic talents had been before, but my newly subpar motor skills had surely undermined them.

"This tree is terrible," I said. "The boys coming over could do a better job."

"No, it's so much better than mine," she fibbed.

Her kind words made me feel better for a while, but inside I sensed that this could be a very long journey. I felt in my gut that something was really wrong with me.

. . .

By Christmas Eve the house seemed to be filled with all the holiday trappings that I had seen on TV—the string of lights lining the exterior of the house, the stockings hung along the fireplace for everyone, including the two dogs, the dishes of green and red M&M's, and the plates of the cookies Taylor and I had made, even the nativity scene with ten-inch figures that covered a tabletop in the foyer.

That night after dinner I was finally feeling strong enough to help with decorating our nine-foot artificial pine tree. Grant came over, and we headed into the garage to bring it down from the attic, where the four sections that made up the main post were stored in boxes. Joan climbed up the ladder and handed down each section to Grant, who passed it on to me. All together, the thing must have weighed three hundred pounds, and it didn't help that the two biggest pieces were stuck together.

Joan explained that we used to have two trees. One was a soft white artificial pine, which we decorated with the perfect red round bulbs, red velvet ribbons, and lights you'd see in Macy's. The other was also an artificial pine, on which we'd hung our family's collection of homemade and sentimental ornaments. After realizing that the pine tree was much more meaningful, we'd done away with the white one entirely.

Grant did most of the assembling in the foyer, extending the branches into position, checking the lights, and helping us flesh out the needles that had been crushed from a year in the attic. The tree began to look elegant as I watched Grant work, and I hoped that I'd taught him to do it that way.

Even though Grant had his own apartment, I found it odd that he hadn't come around much since I'd gotten home. He'd been complaining of stomachaches and not sleeping well, and I wondered what was going on with him.

Joan brought in two bins of bulbs and ornaments, with a gold and red skirt to go underneath the tree. Apparently the coveted job of hanging the big star alternated between the kids each year, and I had the honor of lifting up the winner so he or she could reach the top spot.

"You did it last year," Taylor told Grant.

"Okay, let's go," I said, crouching to grab Taylor around the thighs and hoist her up.

"No, no," Joan said, worried.

"No, I can lift her," I said, and proved as much.

I was amazed at how many ornaments we'd collected, but even more, I was enthralled by the stories that went with them. After Joan and the kids peeled away the tissue paper and revealed each one to me, I watched as they hung them on the tree.

Joan's favorite Kenny Rogers CD of holiday music was playing as Grant showed off the prized paper plate ornament with red and green macaroni that he'd made as a kindergartner.

Some of the store-bought ornaments reflected our children's accomplishments, such as Grant's hockey skates or Taylor's pair of glass ballerina slippers. There were also motorcycles, football players, ice-skaters, and even a trinket to represent every dog we'd owned. I was struck by the little white pillow with the words *Our First Christmas* stitched in red, partly because Joan gripped it so tightly as she told me we'd had it since we got married in 1984.

Our marriage must really be important to her.

When Joan pulled out a miniature stuffed moose on skis that we'd gotten on a family trip to Telluride, Colorado, they all started laughing.

"You want to fill me in?" I asked.

After she explained that I'd had a bad cold on that trip and kept blowing my nose like a moose, I was laughing too.

The cable car ornament, Joan said, was a memento of our trip to San Francisco, where her grandmother Anna lived. Reminiscent of Aunt Bee on *Mayberry R.F.D.*, she was a jolly old German woman who had become like a grandmother to me because my own grandparents had died before I was born.

Grant, antsy to go see his girlfriend, left before we'd finished, but by then I'd noticed the curious number of angel ornaments, including a glass child with wings and a dark-haired child that Taylor had unwrapped so gingerly. Wearing a purple robe, the angel was praying while floating on a cloud.

"This is Taryn's, and we always put it up next to the star," Taylor said softly.

Joan sat me down, and her voice broke as she started to tell me the most important story of all, cupping the family's most treasured ornament with both hands before she handed it over to me.

"I was so hoping your memory would return before we had to relive this. Taryn was our first child, and she was stillborn fullterm. Oh, God, Scott it was awful," she said, pausing as she began to cry. As Joan talked and wept, I stared down at the angel, knowing I was supposed to feel the same loss that she did. But I simply felt blank.

"She was so beautiful," Joan went on. "They called it a cord accident since the umbilical cord was up over her shoulder when she was delivered. Her blood flow was cut off when she lowered into my pelvis and I had contractions, so the cord was pinched off.

It was the worst day of our lives, and we have always agreed that nothing outside of losing Grant, Taylor, or each other will hurt so deeply."

Joan said this incident had reshaped how we viewed the world, and she apologized for making me learn this all over again. "I know this is a lot. I'm sorry, I love you. . . . Do you want to finish the tree?"

I said yes, but the festive mood had been broken. Seeing Joan cry had made Taylor cry, and I'd started crying too. How could I not remember such a traumatic chapter in our lives? It had to be one of those days I'd heard people talk about, the ones that are so important you remember every little detail about them, from where you were to what you were wearing, what you ate, and how things smelled. Joan had lived through those memories, and yet I couldn't even remember my own name or birth date without help, let alone a tragedy that must have hurt me to the core. I wondered if I would ever recover not just the memories but also the pain that came with them, the pain that we'd shared. Watching Joan and Taylor's reactions made me feel even more isolated. I just hoped I'd wake up the next morning and everything would be back to normal.

. . .

The next morning started out like every other one since my accident: it was bright, sunny, and about seventy degrees outside. This morning, after another sleepless night, my headache was still raging. I hadn't been sleeping more than one to three hours since my accident. On a scale of one to ten, the pain was an eleven. Thankfully, the Percocet was slowly working.

After seeing the winter wonderland on television, it didn't seem right that our local trees and streets weren't blanketed with white flakes. "Where's the snow?" I asked, wondering how it smelled and tasted, sure I used to know such things.

"Honey, we live in southern Arizona. We just don't get snow here," Joan said, explaining that we'd often traveled to see snow in

Flagstaff or go skiing in Colorado. Besides, she added, "Christmas isn't about snow. It's about the birth of Christ."

Joan and I had been taking the dogs for walks in the neighborhood over the past few days. Joan said we lived near an area that allowed people to keep animals such as turkeys, llamas, goats, and horses, so she took me to see some. I recognized the horses, but I got really frightened when a scary beast I didn't recognize started running toward me. Luckily, there was a fence to stop him.

"What is that?"

"It's an ostrich," Joan said. "It's a type of bird that doesn't fly."

That made no sense to me, but as I took in the landscape of cacti, palm trees, big boulders, sand, and gravel, I was starting to form a sense of the place I lived and comparing it to Chicago and Buffalo, where I'd seen the people on television, all bundled up and tromping through snow.

I wonder how our Christmas is different from someone who lives in New York.

Compared to all the other stuff I didn't know, this was one of the more pleasant things I wondered about.

. . .

When the kids were little, Joan said, we used to cook a Christmas breakfast of pancakes or homemade waffles, usually with chocolate chips. But now that they were grown, she and I ate while Taylor slept in. She used to cook us pork sausage, she said, but now that we were both eating healthier, it was turkey sausage. After breakfast we sat in the living room enjoying the unusually peaceful stillness of the morning.

Once Taylor got up, she started poking around the mound of gifts that had appeared under the tree overnight. Joan had been wrapping for days. As soon as Grant arrived at noon, we passed them around.

Taylor went first, opening a box to reveal a brown and pink Tory Burch tote bag. Joan had told me that Taylor loved this designer, so it should be a big hit with her. Well, she was right. Taylor was so excited she jumped up and down on the couch, still clutching the bag.

But when it came time for Grant to open his present from Joan and me—a series of gift cards from Chili's, Subway, and Starbucks—he looked irritated and wouldn't look us in the eye. He also rejected Joan's offer of a Christmas cookie.

"You okay?" I asked.

"Yeah, I just don't feel good," he said.

At first I thought he didn't like what we'd gotten him. Then I thought maybe this was how men were supposed to act. I'd been watching him closely for his reaction, so I could determine how I was supposed to respond when I opened my gifts. On *The Sopranos* only the women shrieked with excitement when opening their presents while the men were more laid-back. But Grant seemed so somber, I was confused.

Joan had told me that he'd wanted these cards because he had no money and loved to eat out, but he was clearly displeased. I didn't understand how my two kids could respond in such polar opposite ways. I figured that Grant should be happy getting a gift, no matter what it was, and yet he didn't even seem to want to be there.

It wasn't just that. My whole family seemed to be on edge, which I attributed to my injury. I didn't want them to feel this way. I wanted them to experience the emotional closeness and joy I'd seen on TV, where Christmas was the happiest day of the year.

As Taylor opened up her other gifts—jackets and necklaces from us and a Juicy Couture key ring from Grant—she turned to me and said, "Stop watching me. Open yours!"

So I did. Taylor gave me a spray bottle of Dolce & Gabbana Light Blue cologne. I pulled the blue cap off the opaque white bottle and sniffed. Unsure of what to say, I aimed for a middle-of-the-road mix of pleasure and surprise.

"It smells good," I said. "I like it."

I was relieved that it wasn't something I had to figure out. Frankly, I would have been happy with a cookie because at least I'd know what to do with it.

Joan got me some golf shirts, and I felt bad that I had nothing for her. I knew I couldn't have left the house in my condition, and Joan told me later that I hated going to crowded malls, so I used to pick out and pay for all my Christmas gifts and then get Robyn, my assistant, to pick them up for me. Joan said she hadn't said anything about gifts this year because I had enough to worry about. In the photos I saw of our last few Christmases, Grant never smiled or interacted with the family; he seemed isolated and withdrawn.

Why is he so different from the rest of us? What is going on in his life that he's not happy at this time of year? And why do Joan and Taylor seem more tense when he comes over?

Joan's parents arrived later that afternoon. They'd come over to see me a few days after I'd gotten home from the hospital, and I'd found no resemblance between them and Joan. They also seemed old compared to pictures of my parents, even though they were roughly the same age.

Harvey, a retired salesman, was about five feet five inches and seemed small and frail compared to me. His wife, Fran, looked like the typical TV grandmother—a bit overweight but even shorter than Harvey, with graying dark hair. Before she'd retired at seventy, she'd been a nurse, like Joan, only she'd specialized in drug addiction. She seemed warm and caring.

"How's the pain?" she asked.

"Unbearable," I replied.

Harvey didn't say much, seemingly oblivious to my condition and to our internal family drama. He seemed more concerned with what Joan said was his usual agenda—eating whatever was in front of him and napping on our couch. I did notice, however, that he and Fran seemed comfortable with us and engaged in conversation. It was clear they'd been to our house many times before.

It was interesting for me to observe their moods and body language as they interacted with Joan because I wanted to see how a child, other than my own, acted toward a parent. It was also important for me to watch how Joan treated her parents so I would know what to do when my parents visited.

When my niece Jamie showed up with her family, I tried to put on a happy face, but it was difficult. I was feeling anxious about how I was supposed to act and fearful about their response to me. Grant was still acting distant, even from his little cousins, who jumped on him and tried to get him to play games with this black box that said *Nintendo* and connected to the TV somehow. But he shooed them away, choosing to nod off on the couch instead. I, on the other hand, was fascinated. I thought it was cool that Noah was actually controlling the picture on the TV, and I was impressed that a six-year-old could do so well with his levers and buttons when it looked far too complicated for me to even try.

Surrounded by ten of my closest relatives, I still felt so alone. They were all hugging each other and finishing each other's sentences. I felt like an outsider, as if I had nothing in common with any of them. Even though everyone went out of their way to make me feel like nothing was wrong, I still felt invisible. I wanted to scream, "I'm not okay, and I'm scared!" Despite what the doctors said, there were moments when I was so afraid that the rest of my life was going to feel this empty that I had to fight to hold back the tears.

It helped when Joan and Kevin set me up outside with the natural gas-fueled barbecue and got me started grilling the chicken, steak, and prawns, which provided a welcome distraction. I'd watched a grilling show on the Food Channel, but I had no independent recollection of how to do my usual dinner duty. Thankfully, Kevin suggested when it was time to turn the meat.

When Joan said dinner was ready and called us to the table, everyone started moving toward certain seats, as if they knew where they were supposed to sit.

"Do all people sit in a particular chair?" I asked Joan. "Where do I sit?"

"You sit at the head of the table, right here," she said, guiding me to the end nearest the kitchen. "Grandpa sits at the other end."

. . .

We had sweet corn, Jamie made "her delicious" garlic mashed potatoes, and Fran had brought her famous sweet potatoes in their sticky syrup, with bakery-fresh apple pie for dessert.

It had been a rough day. I was still feeling weak physically, so I couldn't roughhouse with the boys as I apparently used to, but I played with them in short intervals the best I could. I let them climb onto my lap, press their sticky gift bows onto my shirt, and pull my Christmas cracker, enjoying their screeches of glee when it popped and their little toys flew out. I also tried to get to know the rest of my family a little better and even put on the brightly colored tissue paper crown like everyone else. Part of me wanted to join in the celebration, and part of me just wanted to doze off on my chair, but I managed to control my headache with the medication and didn't have to go lie down. That night I fell into bed, my body aching and racked with exhaustion from the most physical exertion I had done since the hospital. Still, I was relieved that I had managed not to scare anybody or make anyone sad, and most important, I hadn't broken into tears. I'd made it through my first family holiday, and for that I felt thankful.

T HE NEXT DAY Joan started taking down all the Christmas decorations inside the house, except for the tree, in preparation for Taylor's birthday on the thirtieth because she wanted Taylor to have her own, separate holiday.

"No combo celebrations," she said.

On the twenty-ninth Grant came over to help strip off the exterior lights. Although the sixty-five-degree weather wasn't anywhere near heat-stroke temperatures, he was sweating and looked noticeably pale.

Some of the bulbs had burned out, and Joan had already bought a new set of lights for the next year, but Grant made the simple removal job harder than it needed to be. As he ripped off the plastic cord in an agitated frenzy, he cursed when the lights didn't come loose easily, then tossed and broke them on the front walkway.

I'd heard people cursing on *The Sopranos* and in movies I'd been watching and asked Joan what some of the words meant.

"*Ass* is another name for butt," she said, warning me before I repeated any of the other words, which she said were generally considered inappropriate. "We try not to swear, especially around the kids."

So, hearing my son spouting expletives that afternoon, I came outside to investigate. "What the hell are you yelling about?" I asked. "God, Grant, you're making a mess."

"What's the difference, we're throwing them out anyway," Grant retorted, continuing to swear as he tore the cord from the roof.

"Well, I would rather get up on the ladder and do it myself than listen to you yell," I snapped. Even in my fragile state, I preferred to risk further injury than let the neighbors hear any more of his tantrum.

"Fine, you can do it then," he said, climbing down and storming off.

With my head pounding, I pulled myself up the steps and heard Joan scolding Grant. "I can't believe you let him climb that ladder when he just had a head injury," she said.

I'd pulled down five feet of lights when Joan made a beeline for me. "What's going on? I could hear you guys yelling."

I filled her in, and she told me to come down. "I will do this," she said. "I don't want you on a ladder."

I knew she was probably right, so I did as instructed. Grant came back outside with a bottle of water and finally did as I asked, unhooking each light from its eyelet and handing the cord down to Joan while I watched, red in the face from the pain and irritation.

"I just don't feel good," Grant said. "I don't want to do this."

When he finished twenty minutes later, he mumbled that he was leaving and sped off in his 1998 Honda Accord.

"I don't know much," I said, "but I wouldn't think I'd let my dad get on a ladder if he'd just gotten out of the hospital. What is with this kid?"

I couldn't understand how or why my only son would act so selfish and uncaring after crying so hard over my injury at the hospital. It just didn't make sense.

. . .

The next day marked Taylor's seventeenth birthday, which started off with another family tradition. Joan said we always served each other breakfast in bed on birthdays, so Joan and I made Taylor pancakes with a glass of chocolate milk and chatted with her in her bedroom while she ate.

Even though I'd been up most of the night again and my head felt like it was in a vise, I was not going to miss participating in this family ritual. I knew from all of Joan's efforts to make this day special for Taylor that celebrating birthdays was something I needed to learn how to do. I wanted to please Joan and Taylor, and I figured I'd better get used to doing things when I was in too much pain to enjoy them. At the same time, I didn't want my pain to distract or take away from Taylor's day. I didn't know the name for this emotion yet, but I was feeling guilty about my accident and how it was affecting my family.

Taylor spent the rest of the morning and afternoon with her boyfriend, Anthony, while Joan and I relaxed and wrapped her presents: a bottle of Juicy Couture perfume and some designer clothes she'd wanted.

That evening Joan and I took Taylor to P.F. Chang's, her favorite Chinese restaurant, where Grant met us and behaved badly.

Joan's parents and Anthony joined us at the house for dessert and the opening of gifts.

Joan had prepared me for what we did on Christmas, but she didn't give me a heads-up about what happened next, and I didn't like surprises. She turned off the lights in the kitchen, where we were sitting around the table, with an ice cream cake in the middle, and everyone started to sing. But there was only one problem: I did not know the words to the birthday song. Feeling very uncomfortable, I watched everyone else and tried to mimic the words by mouthing along. I'm sure I was way off, but I didn't like looking stupid because it was such a basic tune, so I tried to appear as if I was keeping up.

· · ·

I soon learned that I could feel a little less overwhelmed and frustrated at how much I didn't know by actively learning whatever I could in any way that I could. Watching TV seemed the simplest, fastest, and most comprehensive method, and it became like a life-sustaining medication that was just as important as my painkillers, if not more so. It also helped me cope while I suffered from severe insomnia.

In the beginning I'd get so tired that I'd go to bed at 9:00 P.M. Joan lay down with me and rubbed my chest until I fell asleep, then left and came back to bed an hour or so later when she felt tired. At first I was scared when she got into bed with me, and although I wasn't as uncomfortable with her touching me as I'd been in the hospital, I was still feeling uneasy about it. I wondered if I should say something or just let it go, and I decided to choose the latter. I wanted to do everything normally—act the way I used to—and I figured the best way to do that was to follow her lead. It took me a few days to get used to it, and then it was okay. In fact, I grew to enjoy the attention.

As time went on, Joan and I went to bed together around 10:00 or 10:30 P.M., but I was lucky if I fell asleep for an hour or two before waking up and going to my chair in the living room, where I sat up for the rest of the night, flipping around the two-hundred-plus channels on DirecTV. The satellite service offered me a channel for almost any topic I could want, from business or political news to stock market tips, sports, cooking, history, and movies. With my fallback standard, Fox News, starting up at 3:00 A.M., I never had a problem finding something to watch. Some nights I couldn't fall asleep at all, and if I dozed off for an hour in the afternoon, that would be the only sleep I'd get for forty-eight hours.

Often the pain was so bad that I'd have to take my medication before I even lay down, so it became a challenge of timing. The pills took up to forty-five minutes to work, and I had to take them every four hours to focus on anything, including sleep. Frequently I'd have to get by with just resting my eyes while I listened to the TV. Joan didn't tell me this at the time, but the Percocet often made me quite irritable. I couldn't figure this out for myself, of course, because I had no basis for comparison.

The basic knowledge I gained from watching around-the-clock TV—stopping only to sleep, to eat, to talk with Joan or Taylor, or to take the occasional trip out of the house—helped shape my immediate responses to whatever was going on around me. Over time I would come to understand that the world I saw on sitcoms and in movies was far from the real world, but it did allow me to form a basic understanding of our culture.

It was still difficult for me to learn and retain information, so I used any tool that helped me remember things. I often scrambled the days of the week, for example, or forgot the names of certain days.

One day I saw a commercial for the NuvaRing, the once-a-month contraceptive device. Now, I didn't really understand what this was or why they'd advertise a product for such a private purpose on television. But the ad, which featured women ripping off the midsections of their yellow bathing suits and swimming together like they were dancing, did teach me something else. The commercial was annoying, but its catchy tune listed all the days of the week, which helped me remember them in the proper order. I couldn't get this song out of my head for months, which meant I never forgot this information again.

Later, Joan and I were watching a documentary on Olympic training that showed women doing the same kind of swim-dancing, and I said, "Hey, that's just like the NuvaRing commercial."

When she explained that this was called synchronized swimming, which had its own Olympic event, I realized that the ad had been educational in more ways than one.

My headaches continued virtually round the clock, and even though I managed to watch endless TV shows, I never seemed to complete a full program because my attention span was so short. Constantly changing the channel, I could watch a show or movie repeatedly without seeing the same scenes twice.

Joan and the kids encouraged me to watch movies I'd liked before or were family favorites. In turn, I shared with them movies I thought were good, only to be told that we'd already seen them countless times. It was reassuring to know that some of my likes and dislikes had remained the same, that I might not be as different as I felt.

Maybe everything about me hasn't changed. Maybe it just feels that way because it's all new to me.

Wanting to blend in with the people around me and be able to hold an intelligent conversation, I tried to absorb as much information as I could about real life. I figured that everyone was educated about world events, and to prevent anyone from thinking I was ignorant or lacked a general knowledge of these topics, I watched shows on CNN, the Discovery Channel, the History Channel, and the Military Channel. But because I lacked context, it was difficult sometimes for me to differentiate between breaking news and old news footage, like the space shots I saw in the *Apollo 13* movie.

I'd had no preconceived notion about our planet, but I was still surprised to learn that the earth was round. I was also intrigued to hear that so few people had traveled into space and that the universe never ended. It seemed that so little was known about space—or how the brain worked, for that matter—that after watching documentaries on the subject, I figured I knew as much as most anyone else. I also learned about our history of wars and foreign conflicts, feeling surprised and saddened that so many young men my son's age from all over the globe had died, and were still dying, in battle; I'd assumed that most soldiers would be closer to my age. I also didn't understand why many of these wars got started, and all the differing opinions I kept hearing about this didn't help me form one of my own.

Watching the brave soldiers fight for our country made me wish that I'd joined the military. It seemed like such a noble, honorable thing to do because my country was something worth protecting. Joan said she thought I'd been accepted at the Air Force Academy when I was younger, but she wasn't sure. So I asked my mom, and she said that I'd been recruited during my senior year of high school—and even had been endorsed by a congressman—but I'd chosen to go to Northern Illinois University instead.

It made me a little crazy to watch footage of other major historical events, such as the assassination of John F. Kennedy, because I knew I *must* have known these things before, and yet I had no recollection. I still didn't have a good grasp of time, but I could do the math, piecing together that I'd been about a year old when Kennedy was shot.

I felt myself absorbing lots of information from these programs, but I always felt stressed about whether I was gathering the *right* information—and enough of it—to carry on a normal existence outside the cocoon of my house. Also, the more I learned, the more I realized how much I didn't know. I didn't realize, of course, that this is true for many people because the amount of information available to us these days is almost infinite. It's just that most people don't experience as much fear or anxiety as I did from such an epiphany because the gaps in my knowledge and experience were so vast.

But I could go even deeper into the emotional vortex than that. Watching shows such as *The First 48* on A&E, which documents the activities of homicide detectives and crime scene investigators during the crucial first forty-eight hours after a murder is committed, I was amazed at how they put clues together to solve these crimes. And yet I also found myself wishing that I was one of the victims. I figured my family would be better off that way, not having to live with someone who had changed and forgotten so much, someone who didn't know if he'd ever get better or, God forbid, got even worse. It's not that I was feeling suicidal— I didn't even know what suicide was—nor was I worried about becoming a homicide victim; it was simply a potential method of relieving my mental and physical pain. Of course, I never told Joan about these dark thoughts; I didn't want to make *her* pain and anguish any worse.

Things weren't always dark for me, though. Every day I woke up hoping that my memory would come back and I could return to my previous life. The doctors had said this would happen, and Joan and I were trying our best to be optimistic, but we were getting a bit impatient. We wanted to know what was going on in my brain. But that didn't change the fact that I didn't know how to exist in the meantime. Joan told me I'd been a solution-seeking guy before, and that hadn't changed. So, rather than sit hopelessly by and wait, I just kept flipping the remote, which never left my hand. It was something I could do on my own, and I could use what I learned to become more independent, all of which helped me get through the day.

. . .

Photos, with Joan to narrate the backstory, became another important way for me to learn about who I was, what I'd done, and where we'd been together. Joan had started organizing our twenty-seven years' worth of photos together into cardboard photo storage boxes, categorized and divided by month, year, and subject and kept in her home office. Sometimes she'd mention a place like Sea-World, I'd ask what it was, and she'd say, "You want me to show you a photo of when we went there?"

Depending on how bad my headache was and also on how much information I'd already taken in that day, I'd either say "yes" or "not right now."

One day I was going through some stuff in my office when I came across two quilted binders on a shelf.

"What are those?" I asked Joan.

Joan said they contained Grant's and Taylor's early childhood photos and mementos, compiled in a personalized album made by a friend of my mother's. Taryn's was in the safe.

"Do you want to see the baby photos?" she asked.

I agreed, thinking this would be a good way to learn about my kids, my life as a father and husband, and more about the child we'd lost. Joan pulled the two scrapbooks off the shelf, then punched in the combination on the safe to retrieve Taryn's album. She breathed a deep sigh as she pulled it out and held it with great affection as she placed it with the others on the ottoman in front of my big chair.

We sat in my chair together as Joan slowly turned the eight-by-ten-inch pages documenting my children's beginnings. Even though Taryn had been born first, we started with Grant. It was clear from Joan's reaction going through the ornaments, and now this, that Taryn's would be the toughest for her to get through.

The front of Grant's album featured a photo of him as a newborn, wearing a tiny skullcap. The picture was framed with a series of progressively larger rectangles, covered in soft white cotton with a pattern of little pastel-colored numbers and football icons, similar to an athletic shirt, and bordered with blue lace.

Inside, two ultrasound images were mounted on the first page.

"You can see inside where Grant was growing," Joan said, but I was baffled.

As she described how the baby grew in her stomach, I was able to conceptualize the idea, but I still couldn't make out the actual shape of the baby in all those black and gray shadows. I also had a hard time figuring out how the baby came out, although I didn't ask because I was trying to build trust with this woman and I didn't want her to think, "How do you not know this?"

Next came a page with four hospital bracelets, two each for Grant and Joan. For once, I actually knew what these were because I'd worn one myself. There were also photos of a much younger Joan with brown hair, and me standing behind her with my hand on her pregnant belly.

When we got to the photos of Grant's christening, Joan tried to explain in simple terms what that was all about, but religion was somewhat complicated for me to understand, so I just nodded.

Taylor's album was more feminine, covered with bright red cotton and a border of white frilly lace, her name and birth date stitched in red and blue in the center of a heart. It too contained ultrasound images, hospital bracelets, and newborn photos.

I enjoyed watching Joan's face light up as she showed me a photo of her smiling as two-and-a-half-year-old Grant kissed her beach-ball-size stomach. I could see how special and sweet that moment must have been for all of us.

One photo featured me cutting something attached to Taylor's belly in the hospital. "What is that?" I asked. "It looks like a rope."

"That's the umbilical cord," Joan explained. "That's how the baby is connected to the mother and gets its food and oxygen."

"Did I want to do that?"

"Yes, that's what fathers do."

In Taylor's christening photo, she wore a white silk gown as she sat propped up against a pillow on the couch. She was sandwiched between two-and-a-half-year-old Grant, dressed in a suit and tie, and his slightly older cousin, Sydney, both of them holding Taylor's hands. There was also a photo of her white sheet cake, with *Congratulations, Taylor* etched in pink icing.

Finally, on the last page, there was a photo of me, my mouth wide open, laughing, and holding a girl's shoe that was covered in mud. Joan chuckled as she told me the story of how my twelve-year-old niece, Jamie, had gotten stuck in the mud and lost her shoe the day of Taylor's christening while trying to fetch a ball in the back of our house. I'd gone after her, rescued her and the shoe, and ended up covered in mud.

It did us both good to laugh and appreciate the humor of the situation. It made me feel closer to Joan when we could share the same emotion in the present, particularly when it involved something from our past. That said, it was still difficult when Joan started tearing up because I felt that I was supposed to be weeping too, only I usually didn't feel the same level of sadness.

That was not the case as we looked through Taryn's album. We both cried as Joan turned the pages, especially when we got to the photos of our baby girl, her eyes closed and her face and tiny body marked with purple patches, a pattern called lividity that I learned forms when the heart stops pumping and blood pools at the lowest point of gravity. The celluloid pages contained a locket of her hair, her teeny footprints, her birth certificate, and finally two photos of her grave, taken on different trips. One of them showed a bronze vase of fresh sunflowers, petunias, and calla lilies and the other a collection of balloons and flower baskets.

I now had the story of the ornaments and these photos to help put the memory of Taryn together, but I still had no emotional attachment to our firstborn child. And that only made me feel more confused, lost, and empty. I could see Joan's grief, which still seemed so raw, when she talked about our first daughter. I wanted to feel that same pain again because she'd explained how we'd shared it over the years and how it had strengthened the bond between us. Unlike my headaches, this wasn't a pain I would dread or merely endure. This was one I would welcome.

6

A FEW DAYS into the new year, Joan walked the phone into the living room, where I was watching TV and nursing a whopper of a headache, and told me that my friend Phil Herra wanted to say a few words.

We had just received a Christmas card from Phil and his family, picturing his wife, Linda, his two sons and two daughters, and their springer spaniel. Joan had explained that Phil was an old college football teammate and had been one of the four groomsmen at our wedding. She also showed me his photo in our wedding album, which was on the shelf next to the baby scrapbooks. When I'd gone into financial planning after college, he'd spent a year teaching and then went into industrial sales. The Herras were the only couple we'd stayed close friends with since college.

The couple routinely called us after the first of every year to wish us a Happy New Year and a Merry Christmas, but this year was different. Joan had spoken with Linda several times since my accident, keeping her and Phil apprised of my progress and how Joan and I were doing.

After showing Joan my "I really don't want to do this" face, I reluctantly spoke with Phil. Joan had already described him as an enthusiastically vocal man, who at six feet three inches and two hundred and ninety pounds had the type of voice that carried for miles. So I was surprised to hear him on the phone, sounding mild mannered and speaking in a soft, gentle voice, as if he knew this was going to be difficult for me. He also sounded very positive, which I appreciated.

"How are you feeling, Bossy?" Phil asked.

Joan had told me that my NIU teammates had nicknamed me Bossy during my freshman year after I and the five other freshman offensive linemen were told to shave our heads. With my bald scalp, I reminded my teammates of Curly in the Three Stooges movie where he enters a cow-milking contest in a boxing-ring setting, with *K.O. Bossy* printed on the back of his satin robe. The nickname had stuck, apparently.

"I had an accident and hit my head, and I have a bad headache," I replied. "I apologize for not knowing who you are."

Before I handed the phone back to Joan, Phil reassured me that this was a temporary situation from which I'd fully recover.

"If anyone is going to be okay, it's going to be you," he said.

. . .

After hearing about my accident, our friends Randy and Johnna Leach offered to bring over a home-cooked meal so Joan would have one less thing to worry about. On a day when my pain seemed to be under control, we invited them over. Joan and I thought it would be a good idea for me to interact with friends with whom I felt safe.

Beforehand, Joan showed me their photos and told me we'd met them at one of Grant's motocross events in January 2003, where their son, Justin, had been in wheel-to-wheel competition with our son. Joan said our sons were rivals, but we had been able to put that aside because the Leaches were open, friendly, and animated people, and we enjoyed each other's company.

When Randy and Johnna arrived at six o'clock—with Justin as a late addition—I must have greeted them with a blank stare because Johnna mentioned it later in the evening. "When we walked in the door, did you feel a familiarity?" she asked, never one, as she put it, to "ignore the elephant in the room."

"With you guys there's no memory, but there's a feeling of being comfortable," I replied carefully, not wanting to hurt their feelings.

That much was true. I could tell that I was comfortable having them in my house because Joan, my barometer for whether I liked someone, seemed fine.

Johnna had cooked up some cheese manicotti with salad and garlic bread. While we men headed into the dining room and sat down, the women heated up the main dish, set the table, and filled our glasses with Coke and Diet Coke. I didn't know how manicotti would taste, but Joan and Johnna had discussed what I would like in advance. The Percocet often diminished my appetite, so Joan put small portions on my plate.

Randy, who is about six feet tall with salt-and-pepper hair—and, as Johnna likes to say, is strikingly handsome—teased his wife in his southern drawl that the manicotti didn't have enough sauce. Johnna, an attractive blonde who stood five feet five inches in heels, replied, "Kiss my ass."

The way she said this was funny, so I laughed. I'd heard TV characters use this same phrase in a mean way or to start a fight, but she was being playful. I looked over at Joan and was relieved to see that she was laughing too, because this was the kind of banter we usually exchanged—even more so, I was told, in the old days. I'd been noticing lately that the same word or phrase could mean different things in different situations, and sometimes I used the wrong word at the wrong time. Other times I knew what I wanted to say but couldn't retrieve the right word, which I learned was called aphasia.

From this joking around, I could see the evening was off to a good start. The Leaches told me stories about Grant and Justin racing together and how Randy and I always parked our RVs next to each other. Justin, who had a good-looking mix of his mother's and father's features, recounted how we used to go to dinner and a movie after a long day of practice runs at the track, trying to relax before a big race.

They seemed worried that I didn't remember anything they were describing, but that didn't stop them from trying. Johnna was sweet but persistent as she kept looking for a trigger, as if she believed that by asking me "Don't you remember this?" enough times, my memory would suddenly return.

Seeing the silver band on Justin's left hand, I asked, "So, you're married?"

When everyone but me started laughing, Justin said, "No, why?"

"Well, that's a wedding ring isn't it?" I replied.

"No, it's a promise ring," he said.

"Promising what?" I asked, confused.

The others were still laughing, but I didn't understand why. "What's so funny?" I asked.

They told me that I always used to tease Justin about pretty much anything. They explained that his promise ring was a formal symbol that he and his girlfriend were going steady, like a pre-engagement ring. I couldn't understand this reasoning, so I kept asking him why he would wear a wedding ring if he wasn't married. Everyone seemed to think I was teasing him like I used to, but in fact I was simply trying to understand.

We munched on Johnna's chocolate chip-pecan cookies and sipped coffee as the Leaches continued to explore what I knew and what I'd forgotten. We also had an interesting exchange about tattoos, during which Johnna showed me the Christian fish on her foot and the cross and dove on her lower back. When I didn't understand any of the references, she asked some basic questions so she could figure out where to start. Joan had already briefed me that Johnna was a "born-again Christian" and "very spiritual" and tried to explain what that meant. Johnna, she said, talked a lot about how God related to her everyday life, and she also used religion to guide her. Joan, in contrast, kept her belief in God more private.

"Do you know Jesus?" Johnna asked me.

I'd heard the name watching shows and a Catholic mass on TV at Christmas. Remembering Joan's description that the holiday was to celebrate Jesus Christ's birth, I said yes, meaning figuratively speaking.

"Do you remember the Bible?"

"Yes," I said, pretending that I knew exactly what she was talking about, although I'd only heard the word "Bible" on television. That seemed to satisfy her, though, because she didn't prod me to elaborate, which was a relief. "God," "Jesus," and "religion" were all just words to me. It was very difficult for me to grasp such abstract concepts because I couldn't see or touch them.

After chatting for two hours, we decided that I should try to relax for the rest of the evening, so Randy, Justin, and I said our good-byes by the counter between the kitchen and dining area. Later, Joan and I discussed whether Grant would have shown someone else the same compassion as Justin had that night by coming over with his parents to make sure I was all right. We both agreed that he probably wouldn't have.

. . .

Now that I was feeling a little better, Joan was spending more time sitting with me in my big chair, perhaps because she was realizing that my memory loss was lasting longer than we'd expected and it was time to start showing me her love more openly. She would often massage my head and neck, rub my arms and torso, then curl up in my arms and lay her head on my chest. It felt good having her there, and I now liked the way she touched me.

Even though it was a new sensation for me, Joan told me that we'd cuddled like this many times before. "My favorite spot in the world for the past twenty-seven years is with my head on your chest," she said.

As she gently caressed me and gazed at me longingly, I could see what she was thinking: *I know you are in there, Scott. Please come back to me.*

I could also feel the warmth in her touch and the comfort and security she felt when I hugged her back. We'd been kissing a little here and there when my pain medications were working at their peak, and the kisses were becoming more passionate. I'd seen various stages of lovemaking on TV and the movie channels, and I knew that's what husbands and wives did, but I didn't really feel moved to go any further, even as I watched her getting in and out of the shower. Why, I wondered, didn't I feel like I wanted to do that myself?

Joan never said anything, and for now, anyway, she seemed to be satisfied with the kissing and hugging, although I could tell that she truly needed this human interaction. And because I knew my brain injury had taken so much from her, it seemed pretty important to give that to her. I sensed that she needed it to stay strong, to know her husband was still inside there somewhere even though I had changed so much. Even with no basis for comparison, I figured I had to be completely different because I was such a stranger to myself.

"I love you so much," she'd say.

"I love you too," I'd reply.

I had no idea what those words truly meant, but I felt that I was supposed to say them back, as I must have done in the past. That said, I *was* actually feeling some real emotion for her now—I didn't feel so alone when this little bundle of woman was curled up in my lap. In fact, as I began to realize, this was the *only* time I didn't feel alone.

It was hard to watch her weep, though, clearly feeling the loss of the connection we must have shared in the past. I so wanted to remember her and all those intimate moments. How could I not remember this sweet, beautiful woman who was so full of life? She was being so open, sharing all her emotions in her touch, with that soft, warm look of attachment in her eyes. I wanted to rediscover my feelings of love for her so she could see the same warmth in my eyes as I looked back at her, connecting with her and comforting her as she was comforting me. But how?

. . .

Joan was on the computer in her office one day, searching for journal articles on the Internet about head injuries and memory loss, still trying to figure out what was wrong with me.

I sat beside her as she searched, and she showed me how she was using a handy tool called Google. I remembered that Dr. Walker had mentioned this in the hospital when he said he'd looked up my past life in the NFL.

"This is a great way to get information if you're unsure of something or want to know more about it," she said. "It's a good resource to the world."

She walked me through a search using the keyword *amnesia,* cautioning me that not all websites were reputable and that I should take care to differentiate between sound and unproven medical advice.

I was a slow typist and my spelling wasn't good, but the great thing about Google is that it corrects your typos, so once I got the hang of it, I used it constantly. Between Google and TV, I was getting a handle on educating myself, and with every search, I felt myself growing a little more independent.

Early on, I searched the web for more details on TV news stories that had piqued my interest, soaking up data about the economy, the stock market crash, the bank bailout measure, our new president, the Bernie Madoff scandal, and the business world in general. Following the lead of my favorite TV news source, I made FoxNews.com my home page.

I joined Joan in trying to research what might have caused my amnesia and the dark area in my right field of vision, neither of which had improved one iota. Along the way I also Googled myself and found yet another way to learn about my past life as a businessman in a way that was not filtered through Joan's eyes but gave a more neutral and objective perspective from news reporters.

That said, it was ironic that even those news reports were filled with quotes from Joan, who, as my marketing director, had been the spokeswoman for our aviation companies, Legendary Jets, West Jet Management, and West Jet Aircraft. Those stories, in turn, raised more questions that I asked of Joan, who showed me more photos of airplanes, clients, and places we'd flown.

Similarly, I found stories about my college football days, my time in the NFL, and even FamousWhy.com, which listed me as "a famous American football player." Who knew?

I was happy to discover a whole new avenue of finding information, which seemed endless. The world was slowly opening up to me.

. . .

But I still had plenty of questions. I watched Joan put on her makeup and get dressed in the mornings, figuring that if I learned her routine, I would know what mine was supposed to be.

I was puzzled about one thing in particular, however. Why was Joan, who knew enough to find her way around the Internet, so absentminded that she constantly misplaced my pain pills? During my first week home, she'd always bring them to me, even in the middle of the night. But after I started feeling more self-sufficient, I'd get them myself.

When she'd go to the store and I couldn't find the orange plastic pill containers where I thought I'd left them last, I'd have to wait until she came back to get some relief.

"Where are my pain pills?" I'd ask.

She'd tell me they were behind the plant, out of sight, or in the kitchen drawer with all the pots.

"Why did you put them there?" I'd ask, seeing no rhyme or reason in those particular locations.

"I was cleaning," she'd say.

I was confused, wondering if this was normal. After this kept happening, it seemed more than coincidental, but not wanting to make waves, I just took note of it.

However, it also didn't help build trust that she kept leaving the room to make telephone calls. Here I was, struggling to find my place in the house and our family, and she was disappearing to go talk quietly somewhere to who knows who. Was she making plans to leave me? Did she think her job with me was done? Was she making arrangements for someone else to come and take me away?

I started covertly following her into the hallway to listen. That's when I realized she was mostly confiding in her closest girlfriends, Karen and Johnna.

"I don't know what we're going to do," Joan said. "He doesn't know the dogs, he doesn't know his kids, me. Nothing. And I don't know how to take over the business."

Once I realized she wasn't trying to get rid of me, I was relieved. I was also glad that she had people to go to for emotional support when she was frustrated with me, my medical issues, our relationship, or our financial situation. I only wished that I had someone of my own to call because she always seemed happier and better equipped to deal with me once she got off the phone. The only person I really had to talk to about my difficulties was Joan.

But now that I fully understood how much the ramifications of my injury were wearing on her, I tried even harder to hide my emotions and to hold back even more details about what a difficult time I was having. This, however, only fueled my inner turmoil and made me feel increasingly frustrated and ill-tempered. Sometimes it built up so much, I felt like I was going to explode.

One day I teared up as I watched a medical documentary about babies being born, featuring scenes of the father holding his newborn for the first time, crying tears of joy with his wife, and cutting the umbilical cord.

I've been there, bringing children into this world, and I want to remember how special it was.

As I felt a mix of emotions overtaking me, I went outside and sat in the lounge chair facing away from the house so Joan couldn't see me. Hunched over with my elbows on my knees and my face in my hands, I cried for a good five minutes, trying to muffle the sobs so Joan and the neighbors couldn't hear me. This was a moment I didn't want to share with anyone—or burden them with. When I was done, I wiped my hands and eyes with my T-shirt and went back to my chair inside, trying to act as if nothing had happened. These outbursts helped release some of my stress, but the relief was only temporary. My well of darkness seemed bottomless.

For obvious reasons, I never told Joan about those moments although I did catch her having one of her own in the shower one morning. Hearing her sobbing, I came in to check on her.

"Are you okay?" I asked, reaching in and rubbing her back.

"Yeah," she said, trying to hide her face, apparently trying to protect me from seeing her cry just as I'd been trying to protect her. "It's just hard."

Sometimes, though, the rage came over me with such force that I was unable to control myself, and I'd snap at whoever wasn't cooperating with whatever I wanted to do, including the dog.

"Mocha, get outside!" I yelled at our brown Lab. She didn't understand why I was raising my voice to her, so she peed on the kitchen floor, which only complicated the situation. "Stupid dog!"

At that point Joan jumped in, trying to calm poor Mocha—and me—and mopped up the mess on the tile. "Go sit down, Scott. I've got it."

But the dog didn't get treated any worse than Joan, Taylor, and Grant—or the rude car insurance agent I cursed out after he wouldn't listen to my side of the problem.

"That stupid son of a bitch!" I'd mutter loudly, usually after a conversation hadn't gone well or I'd had difficulty communicating with someone.

After I'd calmed down, Joan told me this was one bad habit that hadn't changed since my injury, joking that she wished it had. Although I'd been largely nicer and more compassionate since the accident, she said, I was just as short-tempered and even more intense than before the accident. She said we couldn't be sure, though, if it was me or the pain medication.

I wasn't immediately cognizant about the reasons for these outbursts, but thinking about them afterward, I realized I was feeling tortured about being lost within myself, not knowing if I was ever going to feel right again. The new Scott was battling with the shadow of the old Scott, whom I pictured as lost somewhere in the crevices of my gray matter. I was supposed to be getting better, but I felt I was actually getting worse in the sense that I still had no memories. Not a single one had come back as the doctors had promised. Meanwhile, I could sense my family was waiting, desperately hoping that the memories would return along with the man they had once counted on, the man Joan described as "the ROCK," who seemed to have disappeared into thin air.

The old Scott, Joan said, was a guy who knew what he wanted and would speak his mind, loudly, when the mood struck him, which tended to intimidate people, partly because of his size. I wanted to hear more about this man, whom Joan described as an "alpha male."

"What's an alpha male?" I asked.

"Watch the way people react to Tony Soprano," she said, knowing that *The Sopranos* was still one of my favorite shows.

Telling me to disregard the Mafia ties, extramarital affairs, and violent problem-solving tactics he often employed, she explained that Tony and I had been uncannily similar in terms of our language and mannerisms and our approach to running the household, even down to buying the same car, a Chevy Suburban. In one episode Tony said something like "It may be 2003 outside, but it's still 1950 in this house." Joan said that was true for us too in the old-school way we lived: I'd always brought in most of the money, she'd made decisions about where the kids went to school, and we'd never let Grant and his girlfriend lie kissing on the family room couch.

She also said that although I'd retained Tony's strong family loyalty, I seemed much more sensitive and emotional these days. She didn't seem to be making a judgment about it, but I still wasn't sure I liked the feminine sound of that. I'd heard men on TV say, "Quit acting like a girl," or "You're crying like a woman," and yet that's what I was doing. I wanted to be more like what my impression of a man was, the strong rescuer who put his fist down and solved everyone's problems. But at this point I couldn't even solve my own problems, so I guessed there wasn't much I could do about that. I was slowly coming to grips with the possibility that the old Scott wasn't going to reappear, and this was just the way things were going to be.

W HEN JOAN TOOK ME to see neuro-ophthalmologist
Thomas McPhee, we were hoping to get a more specific
diagnosis for my partial loss of vision and, ideally, a cure.
"Is my sight going to return?" I asked him. "Or can I expect to
wake up someday and all the vision will be gone?"

McPhee said my right pupil looked fine and the retina was intact, but he needed to do some testing before he could give me a
prognosis. He had me lean my face into a machine with goggles
and look at a series of black-dot patterns.

I had no trouble seeing any of the dots with my left eye, but I
could see only those that appeared in the upper two quadrants of
my right eye, with total darkness below the line that went from
four to eight o'clock.

The technician printed up the results for McPhee, which were
no surprise to me because they were just as I'd described my vision impairment. The doctor agreed with Goodell's opinion that
my problem was likely caused by a microscopic hemorrhage in the
optic nerve, and he could suggest only one treatment: an aggressive ten-day trial of prednisone. If the steroids worked to repair the
damage, he said, I should start seeing improvement within those
ten days.

We made a follow-up appointment for a couple of weeks later, but on our way out I told Joan I wasn't very encouraged. "That didn't help much," I said.

"Well, let's give this a shot and see what happens," she replied.

. . .

My skepticism unfortunately was proven out. By the time we were ready to go back, I still couldn't see out of the bottom of my right eye, and although McPhee had already told us the steroids were the only treatment he could recommend, we still hoped he might refer us to a surgeon who could offer us an innovative treatment.

Before going to McPhee's office, we stopped across the street at Scottsdale Healthcare to pick up the records from my hospital stay. This went faster than expected, so with the breeze wafting through the windows of Joan's parked Porsche Boxster, I took the opportunity to relax a bit and get some sun while she sat next to me, reading the medical reports.

I was sitting with my eyes closed, on the verge of dozing off, when Joan startled me. "Son of a bitch, you had opiates in your system," she said.

Rattled, I sat up and shouted, "What the hell does that mean?"

"Wait a minute, let me finish," Joan said, sounding pissed off.

What did I do? What did I screw up now?

I tried to wait patiently until she'd finished reading the page, after which she let out a big sigh. "Okay, that makes sense now," she said.

"Good," I said. "Now, you want to tell me what the hell that was all about?"

Joan said she'd read the report from the second ER doctor first, which showed that I'd had opiates in my system, but that was because the first ER doctor had already given me morphine for the pain.

"Okay," I said, "but what the hell is an opiate?"

Joan explained that an opiate was a narcotic such as morphine, Percocet, or oxycodone, like what I was taking for my headaches.

"So why did you get so upset at first?" I asked, confused.

"Because if you took something for pain that morning, prior to the fall, it would show up in your system, and they would think you fell because of the medication you took," she said.

Not understanding the subtle differences in meaning for the words *drug* and *medication,* as in recreational drugs versus prescription drugs, I mistakenly thought she was saying I had a drug problem. I also jumped to the erroneous conclusion that she thought I'd had drugs in my system before I got to the hospital and that's why I fell. If this was true, it would certainly explain why she kept moving my pain medication around the house.

I tried to digest the contradictions in what I thought she was saying and piece the puzzle together. But I couldn't make sense of it; I was going to have to ask her straight out. I looked at her and asked, very seriously, "Am I a drug addict, and if not, is someone in our family?"

Joan's face went blank. I could tell I hit a nerve, and I braced myself for the answer.

"Why would you ask that question?" she asked.

"Why? Because every time I go to get my pain medicine, it's never in the same place that I leave it. You are constantly moving it on me. Why is that?"

Joan put the papers down, turned toward me, and told me something that made me sit up in shock. "Scott, you are not a drug addict, but Grant is. I have to keep hiding your pain medicine because Grant might take it and get high if he finds it when he comes over."

"Oh, my God," I said, processing Joan's reaction at the same time I was dealing with my own. She looked somewhat relieved to have finally told me, but she clearly had not been prepared for my blunt question.

Part of me also felt relieved—that she hadn't been hiding the pills from me. But that emotion was more than countered by the devastation of learning that my son was a drug addict, not to mention the trauma of having that same emotional wound ripped open once more by discovering that I'd forgotten yet another major problem or life event. It seemed that every time I was able to form a thin new layer of protective skin over that wound, another discovery tore off the fragile scab.

Along with these emotions came sadness and fear because I didn't know enough about addiction to help my son. What I'd learned about addiction came from catching a few episodes of the A&E show *Intervention* and VH1's *Celebrity Rehab* with Dr. Drew. Although I knew this was a bad turn of events, I didn't know how bad. Having little knowledge and no context to quantify the extent of this problem, I felt as if I'd been hit in the head with a fifty-pound sledgehammer. Then, with so many conflicting emotions raging through my already overtaxed brain, I went numb.

The only positive development was that Grant's erratic behavior was starting to make more sense. *Intervention* had taught me that drug addicts and alcoholics act like jerks because they care only about themselves. I started running through the times Grant had behaved badly recently: the episode with the Christmas lights, acting out at Taylor's birthday dinner, the way he kept falling asleep, and his irrational outbursts of anger. During his weekly calls and occasional visits since I'd come home from the hospital, he'd usually talk to Joan for anywhere from twenty to sixty minutes then talk to me for about five. I'd been trying to figure him out and build a relationship, but it was difficult. Once he asked how I was feeling and I said "getting better," he'd switch topics to his job, his girlfriend, or himself. Now that the mystery was solved, I felt as though I could understand him a little better, and this gave me at least a small dose of comfort.

But now, after dealing with this bombshell, I still had to go inside to see McPhee for what I expected would be more bad news about my eye.

Just like the last visit, the doctor put drops in and conducted the same tests. "Has there been any change?" he asked.

"No," I said. "There's been no change for the better and no change for the worse."

McPhee began a long-winded explanation about the eye's complicated physiology and the scarcity of treatment available for my condition. I was in no mood for this—I just wanted him to get to the point—so I interrupted him.

"The treatment didn't work, obviously. Will I ever get my eyesight back in my right eye?"

McPhee explained that some people had much worse vision problems than me, and even though my full sight could return, it was unlikely if, as Dr. Goodell had speculated, the fall had caused a hemorrhage in my optic nerve because such damage was usually permanent. The longer my sight didn't improve, he said, the less likely it was going to.

"Is there anything else that can be done?" I asked.

He reiterated that the prednisone was the only treatment he could recommend, and now that it had proven ineffective, he didn't have anything else to suggest.

"Does this mean I'm considered legally blind?"

He said no, my vision was still good otherwise, so I shouldn't have any problems driving, nor would I need additional glasses. That was all well and good, but I had one more question. Joan had already told me that if my sight didn't return, I wouldn't be able to fly planes anymore, but I wanted to hear this directly from the doctor.

"Does this mean I can never fly again?"

"Unfortunately, you can't fly for hire as a commercial pilot, but I'm sure you could fly as an individual, private pilot," he said.

A damaged optic nerve was about the best diagnosis I was going to get for my vision loss. I'd known in my gut that this was not going to change, and even though it had nothing to do with my memory, I'd been wanting this information, waiting for it so I could get some form of closure.

"Well, I guess it's permanent," I said.

"I'm afraid so," McPhee said.

"That's it," I said to Joan. "I will never see out of my right eye again."

That's when it finally hit me. Joan had told me how much I'd loved the freedom of flying a plane above the clouds to escape the insanity of life on the ground, to fly us to Las Vegas for the weekend or to Palm Springs for our anniversary, to take the kids on a family ski trip, or even to fly my parents in on one of our chartered jets. It was my passion, my life's dream, not to mention an avocation that had made us a lot of money. I hadn't sat in a cockpit since my accident, and it wasn't as if I missed being a pilot, because I couldn't remember what it felt like to fly, but I still felt angry. This was just one more thing that had been taken away from me. First my memory, then my eye, now a career. What the hell was next? Even if I did get my memory back, I still wouldn't be able to fly commercially again.

The negative thoughts started coming at me in a rush, my anxiety mounting into a panic with each new scenario I feared would happen next.

If this is permanent, will my memory loss be as well? Is the rest of my body going to start failing too? Will I start forgetting things I learned since the accident?

I felt like I was losing what little hope and control I had left, and a blanket of hopelessness fell over me. I disintegrated into tears, crying so hard I couldn't even talk.

My life is coming apart, piece by piece, and there is nothing I can do to stop it. I'm on a death-defying roller-coaster ride of misery, only I didn't ask for a ticket and now I can't get off.

McPhee sat quietly and let me vent while Joan hugged and tried to console me, even though she was crying now too. When I was finally able to get a few words out, I apologized to the doctor. "I'm sorry, but it's just been a real difficult road here," I said.

"I understand," he said. "But you'll get used to this to the point where you won't even notice it."

He said I could get a second opinion and repeat the MRI if I liked, but he wanted me to come back in six weeks for a follow-up. As we were leaving, I told Joan that I didn't see any point in coming back only to receive the same prognosis again.

I'm not sure if I grieved my vision loss properly, because as soon as we got back to the car my mind shifted back to the Grant situation, with all the emotions that evoked, and I went on another ten-minute crying jag.

Not only is my vision loss permanent, but my son is addicted to painkillers and cocaine. I wish I could remember, but did I not encourage or support him enough? Was I mean to him? Was this somehow my fault?

Joan rubbed my back and my leg, letting me get all the emotion out, until I was ready to head home. Still, on the drive back, my ruminations continued, as I came up with more questions about possible causes for Grant's condition.

After I was settled in my chair at home and had tried to deaden my headache pain with more medication, Joan and I dived deeper into the issue. It had been a rough day, which made it even more difficult than usual to try to absorb a lot of information, but I felt I needed to take in as much as I could.

"When did his drug use start?" I asked.

Joan said Grant began using drugs in his junior year of high school, but we didn't know until it escalated during his first semester at ASU in the fall of 2007. "He was in school barely a month," she said, when we pulled him out in October and took him to rehab. "He wanted help."

From there, the questions I'd been thinking about on the drive home started pouring out of me: "Is it something I did or didn't do it to make him turn to drugs? Did I not teach him the right way to grow up? Did something bad happen in his life?"

Consumed with doubt that I had been a good father, I listened quietly as Joan told me a story she'd been saving until I was ready to start dealing with the negative issues from my past. Up to this point, I'd been having so much trouble coping with what was in front of me, she knew I couldn't handle anything more.

"Well, he did have a head injury when he was eleven," she said. "And we always questioned whether there were some residual depression issues that might have led to his low self-esteem."

Grant was playing touch football during recess at school, she said, and took an elbow to the head. I was at home that day, and she was working in the recovery room at the Greenbaum Surgery Center at Scottsdale Healthcare–Osborn when she got a call from the school nurse, who was a friend of hers. The nurse said Grant had been hit in the head, was vomiting, and needed to be picked up.

Joan asked me to bring him home to relax, which I did, but once we got there he didn't get any better. In fact, he complained he had the worst headache of his life and continued to throw up.

Our son was in tremendous pain and I was worried, so I called Joan at work and she told me to bring him to the emergency room at Scottsdale Healthcare–Shea, which was only five minutes away. Meanwhile, she called some of her trusted ER nursing colleagues to tell them we were on our way. The nurses met us in the triage area and immediately took him back to see the doctor.

Joan, who had worked at the Shea ER only four months earlier, arrived in her nursing scrubs shortly after we did. By the time she got there, one of her nurse friends had already started his IV and was giving him fluids and medicine to stop the vomiting while they waited for him to be taken up for a CT scan. After the scan they moved him into an acute bay, which meant his condition was growing more serious.

Dr. Paul Francis, a neurosurgeon from Barrow Neurological Institute in Phoenix who had just finished a surgery in the Shea operating room, came bursting into the ER.

"Where is the boy with the bleed?" he asked. "Is he still awake? We need to get him into the OR immediately."

"Let's get a helicopter and get him to Barrow," Joan said.

But the doctor said there wasn't enough time. "He will go from awake to asleep to dead," he said bluntly.

Francis told us that Grant had suffered an epidural bleed in the left temporal region and a skull fracture that had punctured the temporal artery. The doctor reassured us that he could perform the surgery at Shea, then have Grant transported to Barrow once he stabilized.

We agreed, so Grant was wheeled up to the operating room, where we stayed by his side until it was time to go in and repair the damage. Because Joan was a hospital employee, she was able to see Grant directly after he came out of the hour-long surgery. She was concerned about his mental functioning, but as soon as he asked about the catheter—"What is in my penis and how did it get there?"—she knew he was going to be fine.

Thankfully, she told me, our son didn't suffer any permanent damage or experience any seizures. As the doctor had promised, Grant was transferred later that day to Barrow and spent the next couple of days there healing. He was released without any complications, but he wasn't allowed to participate in any sports or other activities that might reinjure his head.

This began to take a toll on Grant, Joan said, because he'd been playing hockey on a traveling team in Phoenix and now was forced to sit on the sidelines. "This was the kid who was always picked first to be on the team, the best athlete, almost like a mini-you," she told me. "He was a little alpha male, and for the first time he was stripped of being that kid who could do anything.

"It was like a surgeon losing his hands," she went on. "An adult could have handled it, but an eleven-year-old couldn't understand it. His whole life was, 'I'm a hockey player.' He couldn't go to recess, he couldn't play any sports, and he felt isolated."

I feel like that too.

As Joan went on, I felt a connection growing with my boy for the first time. Maybe this common experience could help me reach out and build a closer relationship with him. And Joan didn't have to cue me this time. I saw it all for myself.

His friends didn't want to play with him anymore; I don't have any real friends other than Joan and Taylor. He must have had pretty bad headaches too.

Also, for the first time in his life, Joan said, Grant was being teased. His head had been shaved for the operation, and until his hair grew in, the long red and swollen scar, configured like the number 7, was exposed on the side of his head. "It was so difficult for him," she said.

When the anger and depression seemed to last longer than they should, Joan said, we decided to seek counseling for Grant and guide him through the process. After months of therapy, he seemed to get better at controlling his temper, but it was not until he was able to return to sports six months after the accident that we saw any real improvement.

By then, however, his team had moved on without him. He was no longer the star of the traveling team; in fact, he'd lost his position altogether. As a result he started pursuing extreme and unsafe sporting activities such as jumping his bike off rickety ramps or off the second-story patio and into the pool. At that point, Joan said, we introduced him to motocross, in which he could wear a good helmet and safety gear and be protected during his "need for speed" phase, which ended up lasting six more years.

Joan's story took about an hour to tell, and by the end of it I was exhausted. I had long ago reached my saturation point for information, even on this very important topic. "I need to relax and close my eyes," I told her. "It's just so much."

As this new reality sank in over the next few days, I began doing some research myself. I found some statistics showing that, at least in some cases, head trauma and drug addiction appeared to be correlated. I couldn't help but wonder if this injury had led to Grant's drug problem.

IT HAD BEEN about six weeks since the accident, and I was starting to go stir-crazy at home, so Joan and I decided it was time for me to go someplace other than the doctor's office. I wanted to contribute to the household, and although I wasn't sure exactly how, a trip to the grocery store seemed like a good start. Still, just the thought of completing this simple task, which most people took for granted, filled me with dread before we even pulled up to the Safeway near our house.

Although I'd seen supermarkets on television, I still wondered what it would look like inside. Would I be able to tell I was in a grocery store? Or, worse yet, would I see people I used to know and have to deal with the awkward situation of not recognizing them?

I gripped Joan's hand tightly, but she had to break away to pull a cart from a long row of them. Right away the place seemed unfamiliar, and I felt lost amid the stacks of cans and boxes surrounding me. I started to push the cart as if I were trying to help Joan, when in fact I was really just trying to stay close to her, my trusted human security blanket.

We passed a display with bouquets of flowers near the entrance and started walking up and down the aisles, where I watched her take items from the shelves and toss them into the cart. I'd never seen most of the items she was choosing, and I was feeling overstimulated with being in a new environment. I wanted to be of help, but I also felt I needed to venture off and clear my head.

"Is there anything I can grab?" I asked.

"Why don't you go get a bag of potato chips from the next aisle?" she suggested, as if that should be an easy task for me to handle.

The problem was I had no idea what a bag of potato chips looked like, though surely, I told myself, I would recognize one when I saw it. Well, I found the correct aisle all right, but I wasn't anticipating there would be more than fifty different types, sizes, and brands, all of which said "Potato Chips." Not knowing what else to do, I started pacing back and forth, scanning each bag and storing the label in my empty memory bank to try to properly evaluate each one. *Barbecue, salt and vinegar, cheddar, onion, garlic, ruffled, baked, and low fat.* If I didn't know what any of these tasted like, how was I supposed to choose? Feeling so unequipped to make such a simple decision made it even more difficult to do so. Beads of sweat formed on my forehead as I told myself that Joan was counting on me to make the right choice and I was going to fail. Just then she came around the corner.

"Honey, just grab a bag," she said, as if there was no wrong choice.

"How do you choose?"

"Just choose whichever one you want," she said.

I grabbed the red, white, and blue jumbo bag of chips with the ridges, called Ruffles, because I liked the colors on the bag and the look of the chips' rippled edges. I had no idea that it was a brand we usually didn't choose, but I did notice that Joan didn't correct me. Only later did she tell me that we usually bought the healthier baked chips these days. I felt dejected. Everyone else on that aisle had made his or her choice so easily. The trip was just one more reminder that I didn't know anything about the outside world.

While we were waiting in the checkout line, I walked over and grabbed a cellophane-wrapped dozen red roses from the flower display. I'd seen Anthony buy these for Taylor on her birthday, and I'd seen plenty of men on TV do the same for their wives or girlfriends.

Joan had been doing so much for me, I thought it would be nice to give her these as a token of my appreciation.

As I handed them to her, she teared up. "Thank you, I love you," she said, giving me a hug and a kiss right there in line. That's when I realized it didn't take much to please Joan. My small gesture had cost only ten dollars.

. . .

Now that I was feeling a little better, I wanted to be more helpful around the house. Joan showed me how to pay our bills online, which I then started to do on my own. I knew from overhearing her conversations that she was worried about money, but I didn't like it when she told me I'd used the wrong credit card to pay a bill that charged no interest or penalty before paying one that did or when she said we couldn't buy something that I wanted her to have.

We'd be in a department store, for example, and she'd see an item that she liked for herself or Taylor.

"I want those shoes," she'd say.

"Well, then, just buy them," I'd say.

"We don't want to spend the money now because we're still try- ing to figure things out," she'd reply.

She'd already told me that I'd been the primary provider for our family, and from watching *The Sopranos* I knew that the man of the house was supposed to say these things to his wife, so this made me feel like less of a man. Worthless. Stripped of the power I used to have, lacking my own identity, and unable to provide for my family like I used to. I didn't think Joan was trying to make me feel this way, but at times her remarks were hard to take.

Then one day I overheard her talking on the phone to Anita, our bookkeeper. Of course, I could hear only one side of the conversa- tion, but I tried to piece things together.

"Anita, relax, we'll get it figured out," Joan said. "We're not in panic mode yet. You need to calm down and get a grip here."

After she hung up I came in to find out what was going on.

"What was that all about?" I asked.

Joan broke down crying as she explained that Anita was concerned that our bills were piling up but no new money was coming in. She'd asked Joan how we wanted her to handle the situation, and Joan tried to tell her that we were waiting to see what happened with my memory, hoping that I could get back to work one day soon. Joan told me I'd been close to finalizing a deal, but she couldn't find any of the paperwork because I'd apparently kept all the details in my head. Meanwhile, she was trying to make some sales by following up with other potential clients.

I didn't know how to respond—I obviously couldn't retrieve the details of that deal now—but I tried to do what I could to reassure her. "We'll get it figured out," I said. "We'll be okay."

. . .

Our financial troubles, not to mention the sad state of the economy, caused me to reflect even more on the material objects I saw around the house. I wondered not only why we needed all these expensive baubles but also whether we should start selling them—my watches and some of the cars, for a start. I also noticed that Joan wore a very nice diamond ring on her wedding finger and a beautiful diamond tennis bracelet on her right wrist.

"Tell me about this ring and bracelet," I said.

Joan was only too happy to tell me about them, saying that we'd had a custom jewelry designer make the ring several years ago. I'd already replaced the small diamond I'd bought for our engagement with a 1.5-carat marquis diamond some years ago. More recently, we'd had that same diamond placed into a new platinum setting with twelve new smaller diamonds, six on either side.

She said I'd purchased the tennis bracelet, a string of six carats' worth of diamonds that shined so brilliantly, to congratulate her for completing her master's degree of science leadership at Grand Canyon University. I was amazed at how proud she sounded as she told me about the gift, which I had given her on graduation day when she was still wearing her cap and gown.

"You are the least selfish person that I've ever known, and you've always been happier when you buy something nice for someone in the family than when you buy something for yourself," she said.

It was wonderful to hear how much the bracelet meant to her, but I couldn't help wondering if I would ever be in a financial position again to show my love in the same way—by purchasing a special gift for the woman who meant the most to me.

. . .

In the third week of January Joan and I went to St. Joseph's Hospital and Medical Center in Phoenix for an EEG, a test that measures the brain's electrical activity, to see if we could get some answers. Rather than follow up with Goodell from the hospital, we'd decided to seek a neurologist at Barrow, because it specialized in brain injuries. Dr. Terry Fife, whose daughter was on Taylor's cheerleading team, ordered the EEG and also referred me to a memory specialist and a neuropsychologist for more tests. Although I still hoped that my memory would come back, I was starting to prepare myself for the possibility that it wouldn't.

As if I didn't have enough problems already, the clerk at the check-in desk called me over a few minutes after she entered the information from my medical insurance card into the computer. "Your coverage has expired," she said.

Joan and I looked at each other, shocked.

"There's no way," I said. "That can't be right. Are you sure?" This didn't make sense because we'd just gotten a prescription filled.

The clerk said yes, unfortunately, but we could pay for the $895 test in cash, if we wanted, and submit the bill to the insurance company for reimbursement.

"No," Joan said, "we're going to have to figure this out. We'll reschedule and come back."

Joan and I went out to the car in the parking garage, where she tried to reach the people who had bought West Jet Aircraft (WJA) back in February 2008. Joan said she clearly remembered that they were supposed to have paid our premiums through February as part of the deal. It was late in the day and these people lived in Florida, so we knew it was unlikely we'd reach them, but we waited for half an hour, hoping they'd respond to our voicemail. We finally gave up and went home to search for the sale paperwork in my office, where we found proof in my desk drawer that Joan was correct.

. . .

That night Joan mentioned that an insurance broker named Jerry Pinto had set up our group health insurance plan when we still owned WJA, and we should give him a call. Joan had mentioned Jerry earlier when explaining the concept of "best friends" and said that he and I had been close for twenty years, ever since my financial planning days in Chicago. She'd also mentioned a guy named Mark Hyman, who lived in Scottsdale, and my college teammates, Phil Herra and Brendan Dolan, who lived in Chicago but whom I didn't talk to as often. Jerry, Phil, and Mark had called since my accident, she said, and so had my cousin Brad, but with everything that had been going on, Brad and Phil were the only ones she'd called back. I hadn't wanted to talk to anyone, so she'd been saving all the important phone messages for when I felt well enough to listen to them.

Now that I had a pressing reason to talk to Jerry, I decided to give him a call, but I was dreading it because I still didn't feel I could carry on an intelligent conversation with anyone. I would have preferred to simply say, "Joan needs your help resolving this insurance problem," and hand the phone over to her.

When Jerry answered, I spoke in the slow and deliberate phone manner I'd unknowingly been using since the accident. "Hello, Jerry Pinto, this is Scott Bolzan."

Unaware that I wasn't being my usual teasing self, Jerry replied in a similarly slow and deliberate voice. "Hello, Scott Bolzan, this is Jerry Pinto. How can I help you?"

The old Scott, I'm told, would have responded with equal sarcasm, but the new Scott was taken aback by Jerry's tone. When I briefly told him I'd had an accident and lost my memory, Jerry was immediately apologetic. "Scotto, I'm really sorry. What the hell happened?"

I gave him a few more details, explaining that I didn't remember him. There was a long pause as Jerry digested the news.

"Well, we're best friends and we always will be," he said.

"Yeah, Joan mentioned that," I said. "That's why I'm calling."

"Do you want me to come out there?" he asked. "I'll hop on a plane if there's anything at all I can do for you."

"No, not at this point," I said.

That was a nice gesture. We must have been good friends if he's willing to fly here on a moment's notice.

Antsy to get off the phone, I tried to wrap it up, asking Jerry to help Joan resolve our insurance problems while my brain healed. "It would mean a lot to me if you took good care of her," I said, happily turning the phone over to Joan.

Jerry said he'd get in touch with the people who bought my company, and within a day or two he and Joan were able to persuade the buyers to pay that month's premium.

. . .

Once we got the insurance sorted out, we went back for my EEG. The technician attached a bunch of electrodes on suction cups to my scalp, and even though I had a headache at the time, he said I had to lie flat.

"How long is this test going to take?" I asked, grimacing inside.

"Twenty-five minutes," he said, setting a timer and leaving the room while I lay there, practically counting the minutes, in pain.

When the results came back, it was the same story all over again—they couldn't detect any abnormal activity. It seemed that no one could tell us what was wrong with my brain, why my memories still had not returned, and why this relentless pain would not let up. *When*, I wondered in frustration, *will these doctors do a test that will actually diagnose my problem so they can treat me and let me return to my normal life?*

In addition to this frustration, I was also dealing with the constant annoyance of tripping over things I couldn't see with my right eye. This often happened if I turned a corner to the right, so I started having to remember to look down more when I walked. *I've got enough to deal with, and now this.*

The new owners let our insurance lapse again the following month, but thankfully, Jerry was able to get us new group insurance through Legendary Jets for $1,200 a month. Meanwhile, with me still clueless about how to run my business, Joan had to start looking for a new job. We not only needed cheaper insurance, our finances were in a shambles.

· · ·

Determined to be more independent and also to overcome my anxiety issues with the grocery store, I decided to take a spin over there by myself in my 2007 BMW 750Li.

I took the set of keys from the center console and put the flat plastic one into a horizontal slot in the dash as I'd seen Joan do, only the lights on the dash came on but the car didn't start. After all this time driving with my wife, I realized that I must not have actually watched her start the car. I sat puzzled at my inability to determine what I was doing wrong. Reluctantly, I had to go inside and ask her for help. "Could you come out here with me?"

"Is everything okay?" she asked worriedly as she followed me outside.

"Just one problem. How the hell do you start this?"

I could tell by her expression that she was nervous about my getting behind the wheel. I knew what she was thinking: *If you can't start the car, how are you going to drive it?* After she showed me what to do—put my foot on the brake and simultaneously push the Start button on the dash—I thanked her and shut the door before she could stop me from going.

I didn't realize how much of a problem my vision deficiency was going to be until I tried to back the vehicle out of the single-car door of our three-car garage. The vehicle had only four inches of clearance on either side, so my lack of peripheral vision to the right posed a challenge.

I backed out slowly and, realizing that I was cutting it too close, had to pull forward to straighten out and try again. I could see Joan standing in the doorway, watching to make sure I didn't hit anything, but I was determined to do this right. I finally made it out, turned the car in the driveway, and headed out to the street.

It felt weird driving this car, knowing that I had driven it hundreds of times and yet feeling as if it was the first time. But even weirder—and a welcome surprise—was that I had retained my procedural memory. Maybe it was in a different part of my brain. I still knew I was supposed to hit the brakes to stop and the accelerator to go forward. I knew that at a stoplight, green meant go and red meant stop. I also knew what a stop sign was. Even so, the brakes seemed very touchy, and because Joan had set the seat and the mirrors to fit her small frame, I had to pull over just past our driveway to make some adjustments and settle down for a moment or two.

You can do this. Just relax. Take your time. There's not a lot of traffic. I wish I could have kept the memories of my wife instead of remembering how to drive.

The center of the dash displayed a navigation screen for the GPS system, which I had watched Joan use a number of times. I figured it would come in handy in due time, once I learned how to use it. But I also knew I didn't need it for this short jaunt because the supermarket was so close—a simple right turn at the end of our street, another right turn, then a straight shot for about half a mile. Still, in an abundance of caution, I decided to follow the same routine from my last trip there with Joan.

I drove about forty miles per hour to the store, five miles under the speed limit, and parked the car in the exact same spot. Taking a deep breath, I walked inside, grabbed a cart, and examined the list of items at the end of each aisle to get an overall sense of the layout for next time.

When I finally made it to the cereal section, I picked up two boxes of Quaker Oats Granola. Joan said I liked the stuff; she wrote the name down for me and even told me where to find it on the top shelf. I was sick of eating oatmeal every morning.

After ten minutes of roaming around with my cart, I still had no more than the cereal. I got in the checkout line behind the other shoppers, whose carts were full, and when it was my turn I put my two boxes on the conveyor belt.

"Why didn't you use the fifteen-items-or-less line?" the checkout girl asked.

I didn't have a suitable answer for her, so I just shrugged.

"Do you have your Safeway card?" she asked.

"No," I said.

"Punch in your phone number," she offered, unaware, obviously, that I didn't remember that either.

"You know what? Forget about that," I said. "Let's just ring it up."

I was a little embarrassed and felt like I was in one of Taylor's favorite movies, which we had watched recently, *Baby's Day Out,* where the baby hero heads out for the first time alone into the real world. But all in all, I had done okay for myself. Other than choosing the wrong checkout lane, I had made a successful trip to the grocery store.

When I arrived home safely, I felt as if I had just climbed Mount Everest. Joan could probably tell how proud of myself I was because she seemed happy for me too.

"You made it back," she said, smiling with obvious relief.

"Of course I did," I said. "Even I could do this."

That said, I didn't know if Joan was happier that I'd managed to find my way home or that I'd managed not to hit the garage door as I pulled in.

O N JANUARY 21, 2009, exactly a month after my release from the hospital, I sat in my big chair with a whopper of a headache, nervously awaiting my parents' arrival. Much like a soon-to-be-adopted child, I wondered if I was going be accepted and vice versa. I clutched my parents' picture and studied their faces so they wouldn't seem like such strangers when I opened the door.

Alice and Lou Bolzan were seventy-two and seventy-four and felt they were too old to travel alone, so my sister Bonnie, who is six years older than me and three years older than our sister Candi, was coming with them from Chicago. They had been planning to come out in May for Taylor's graduation, but when my memory wasn't returning as the doctors had predicted, they decided to visit now. My mother, whom Joan described as an eternal optimist who never wanted to hear anything negative, didn't want to wait four more months to look me in the eye and make sure I was okay.

Joan had given me a bit of history about the relationship between me and my sisters, cautioning that she was conveying my past feelings about them without adding her own perspective. If and when my memory came back, she said, she didn't want me thinking she'd been trying to fill my head with *her* views. There wasn't any bad blood between my sisters and me, she explained, although in the past Bonnie and I had experienced some painful personal disagreements. But I was never really that close with either of them because we didn't have much in common.

Around 2:00 P.M. the doorbell rang.

"They're here," Joan called out.

I let out a big sigh and hoped the visit would go smoothly. It was important to me at this point in rediscovering myself to get a sense of who I was from someone other than my wife and kids.

"Scott!" my mother shouted as we opened the door, lunging to embrace me and kiss me on the cheek. I reciprocated by doing the same.

My father, who was slightly taller than me at six feet five inches (about an inch shorter than he used to be) and about two hundred twenty pounds, came toward me. At a loss for how male family members were supposed to welcome each other, I stuck my hand out awkwardly. When he came in for a hug, I hugged him back.

"Hey, you're looking great," he said, pulling away to look at me. "How are you? It's good to see you."

I instantly felt connected to them. I'm not sure if this was because Joan had told me I would feel the same bond that my children shared with me or if it was something instinctual, built into my genetic makeup, that let me know, deep down, that I'd been here with them before.

I was surprised to see that my dad looked frailer than in the photo and that he walked slowly. I was expecting someone with a more muscular build, like mine, partly because I'd forgotten that people shrink with age. My full-figured mother, in contrast, looked sprightly for her age and much younger than my father.

Bonnie, who was chunky with short dark hair, gave me a hug and kiss like my mother, only it didn't feel the same, and neither did I. Her hug wasn't as warm, and I didn't sense the same kind of connection; I also felt guarded as I had during the hospital visit with my NFL acquaintance. With Bonnie, it almost felt as if I was meeting a stranger whom I'd rather had stayed in the car. This puzzled me at first. I realized that I hadn't studied a picture of her before the visit and vaguely remembered Joan pointing her out in family photos, but based on what Joan had said, I figured it was probably something more complex. Nonetheless, my uneasiness took away from the comfort I felt with my parents.

We sat down to visit in the family room, with me in my chair, my mom on the couch closest to me, and my father next to her. Bonnie sat quietly some distance away. I wasn't sure if she was trying to give my parents more time alone with me or if she was preoccupied with her own thoughts, but I sensed that she felt just as uncomfortable as I did. When she asked how I was feeling, it seemed forced.

In contrast, both of my parents seemed genuinely concerned about my health and asked lots of questions, telling me to lie down if I needed to. "How are your headaches?" my mother asked. "Are you getting any fewer?"

"No," I said. "They're pretty much constant all the time, but I try to control them with the pain medication."

We made small talk for a while, chatting about their flight and the Thunderbird Hotel, where they were staying in Scottsdale. Because I couldn't engage in conversation for long without closing my eyes for fifteen minutes, they let me rest while they puttered around in the kitchen or watched TV.

After my little shut-eye, my parents showed me the scrapbooks they'd brought with them, the pages tattered, with yellowed tape holding down photos and news clippings that were faded and crumbling after thirty-five years. As I flipped through the years, I saw myself as a little boy, in my teen years with surprisingly long hair, almost touching my shoulders, and throughout my NFL career. I could see my adult self in that happy kid's face, unlike Grant, who looked nothing now like he did as a child.

Yeah, that's me.

My father had taken a majority of the photos, and he seemed to remember every sporting event down to the plays, the touchdowns I made, and how I'd plowed through the other players when I was quarterback. Seeing myself in a football uniform at such a young age told me I'd loved that sport from way back.

"Whose idea was it for me to start playing?" I asked.

"It was your idea," my dad said, adding that I used to watch kids playing in the park even before I was old enough to join Pop Warner. "But you couldn't play until you were in the fifth grade. You wanted to start playing the year before that."

Although the album had photos of me playing Little League, at that point I knew very little about baseball because the season hadn't started yet and there were no games on TV. I knew even less about wrestling other than I thought my uniform was strange and tight, with weird headgear.

The love and care that had gone into collecting these mementos was obvious; I heard it in my parents' voices, and I could see it in the white ribbons and buttons that read, That's My Boy, which my mother had attached to several of the high school photos. She'd also painstakingly recorded scores for each football game and enclosed every congratulatory note and letter I'd received, even my letter from the U.S. Air Force Academy thanking me for my interest in attending. Apparently my aspiration to be a pilot had taken root early on as well.

Taking turns narrating, my parents explained the relevance of each photo or news story as they turned the pages. My dad proudly described taking me to football practice, noting that he and my mom had never missed a game. "I just loved watching you play football," he said.

During the California Bowl in college, he said with his eyes sparkling, he was on the field taking pictures because he'd talked my coach into getting him a press pass. I was amazed and yet flummoxed that this man in his seventies could recall the minutiae of my life when I couldn't remember a single thing.

Still, I couldn't believe or understand why they'd kept this stuff for so long. "Why did you save all of this?" I asked.

"Because that's what mothers do," my mom said, as if she were stating the obvious. "I wanted you to be able to share this with your children and grandchildren. And thank God we did!"

The more stories they told me, the more my head hurt, so I had to stop after a while even though I wanted to hear more.

Around five o'clock we decided to get some dinner. I didn't feel up to going out, so Joan got some Mexican food from Nando's, the restaurant where Taylor worked as a hostess.

When we sat down to eat, Joan tried explaining my condition in a bit more depth; we were waiting and hoping, she said, for my lost memories to return.

My mother, however, didn't want to accept that I had changed one iota.

"Scott, I don't see any difference in you," she said. "When I opened that door and I saw you, you looked the same. You act the same as you always have, your sense of humor is the same, and your personality is the same except that you're a little calmer."

"Really?" I asked.

I wanted to believe her, but inside I knew that wasn't possible. How could that be if I couldn't even remember who I'd been? I also didn't feel that I had much of a sense of humor, even though I did make my mother laugh when I joked about noises coming from the bathroom while my father was inside.

My mom was concerned about my pain, which she could see in my eyes, but she kept reassuring me that everything was going to be okay. I think I was the only one who realized that my condition might not go away.

"Scott, this is probably a temporary thing," she said, "but you have to be patient. There are two things you can do—you can laugh or you can cry. Doom-and-gloom doesn't get you anywhere."

She said she had a problem with some people's woe-is-me attitude, but I'd always been mentally and physically tough enough to overcome challenges. Still, I wanted to make her understand that my condition wasn't easy to suffer through.

"It's so hard because I have so many things I have to learn," I said.

"Keep looking forward," she said. "You being who you are, you will be able to accomplish all of this."

Although I wasn't as optimistic as she was, I still appreciated her maternal, albeit Pollyannaish, efforts to try to inspire me. I could tell that neither of my parents realized how difficult it was for me just to get through the day, but I didn't push the issue because I didn't want to worry them or ruin their vacation.

. . .

A few days later my parents, Bonnie, and her daughter, Sydney, whom I was meeting again for the first time, came over to visit with me and attend my nephew Aden's second birthday party. Jamie filled the house with Aden's favorite blue and green balloons and brought party hats that said "Happy Birthday." Grant, Taylor, and Anthony joined us as well.

I had yet to hear about Sydney and was quite surprised to see her show up at my house. I felt a little uneasy because I didn't have enough information about her—other than that she was a product of Bonnie's second marriage—to make me feel comfortable.

I enjoyed having Aden and Noah around because they provided me with endless entertainment and helped distract me from my pain and the feeling of being lost. I felt at ease with the boys. They felt the closest to me in mental age, so I related to them better than most adults. Talking to adults was more like work or being interviewed, but with the boys I could just play and relax. They didn't judge me, they didn't care if I was sick or had a headache, and they didn't treat me any differently. They were also the only two people around whom I felt smarter. Still, as Aden was bragging to everyone that he was now two, I felt as if he was actually older than me.

This being the last night before my parents headed back to the freezing temperatures of Chicago, I felt sad. I didn't know when I would see them again or if I'd remember them the next time.

The boys ran around, squealing as they chased a ball and the dogs around the yard, while I spent some time alone with my dad on the back patio. Even though my head was aching, I was eager to do another question-and-answer session.

"Tell me some stories," I said, asking how he was raised and how he raised me and my sisters. If I could learn who he was as a man and a father, I figured I could learn more about myself and how to father my own children.

Sometimes I found myself staring at him because, as his son, I knew I had to be a lot like him. Joan said we looked alike, had the same work ethic, and were both good family men, but I needed to see that for myself. In the back of my mind I was also struggling with the logic that if I took after my father, then Grant took after me, and with the way he'd been acting lately, that thought didn't sit well.

"Why did you have kids?" I asked.

"We wanted to raise a family and share our lives with our children," he replied.

Watching the boys racing around, I asked, "Was I a lot like Noah and Aden? Those kids never sit still!"

"Yes," my dad said, "you were fast." He recalled that I came home from school one day in such a hurry that I ran up the stairs, tripped, fell and hit my head, and had to get stitches.

One Easter, when I was four years old, he said, my mother dressed me up in a new suit. He wanted to take me for a walk, and my mom told him to hang on to me. Things were fine until our next-door neighbor, George, came outside. Excited, I broke away from my dad and ran toward him. On my way, however, I fell down and ripped the knees of my new suit. Needless to say, my mother was not pleased.

My dad said he used to take me to a barbershop, saying, "If you're real nice, and let the barber cut your hair, I'll take you across the street to the tavern and you can play pool." Apparently I was a terror at the barber's, and this was an incentive. I was too short to actually play pool at the time, but he'd let me toss the balls into the holes.

"You were a good kid," he said.

"What was your job?" I asked.

"I worked at General Mills. I was a millwright, like a mechanic, for thirty-one years."

"Did we go places?"

"We went to Florida almost every year," he said, adding that the trips started when I was thirteen and I came down with chicken pox on our very first venture.

My mother joined us outside and told a few stories of her own. Like my father, she often started off saying, "Scott, do you remember . . ." before giving me the details. I'd let them finish then say, "No, I don't remember that. I don't remember anything before December 17." They kept forgetting, and I had to keep reminding them, but in my condition I couldn't complain.

As a little boy, my mom said, I constantly exhibited spurts of energy, getting into everything. I was so active they had to double-lock the front door so I couldn't open it on my own. "We used to call you Flash," she said.

She said I was also a jokester, even at age three or four. We were on a family outing one afternoon, waiting at the river for the bridge to come down so we could drive across. Candi and I were sitting in the backseat when she asked my dad why the foghorn was blasting. I replied matter-of-factly, "Candi, are you dumb? They do that to tell the frogs to get out of the way!"

My mom said I used to hide my sisters' books in the morning because I didn't want them to go to school without me, and I'd hide their cookies in my toy box, forcing my sisters to search for them.

"What did I want to be when I grew up?"

"You wanted to be an astronaut," my father said.

"No," my mother argued. "He always wanted to play football."

After five or six hours playing Happy Birthday games with the boys and talking with my parents, it was time to say good-bye. My parents were flying out the next morning, and I could tell they didn't want to leave with me still feeling so bad, but they were coming back in a few months.

"I'm so happy that I came because now I can go home and know what I saw, that this is Scott," my mother said. "You're everything that you were before."

I wasn't ready for them to leave either. I still didn't have all the answers I'd been seeking for how to be a good parent, but I realized there were no shortcuts for that. Nor did I feel I had enough information to know who I was as a son, and I'd only gotten a glimpse of what I was like as a child, an adolescent, and a teenager. I'd been hoping that they could tell me who I was as a man too, but sadly, I was slowly realizing that no one was going to be able to define that for me. No one but me.

10

Toward the end of January I was growing frustrated. The doctors had said my memory should return in days, weeks, or even a month, and yet not a single glimpse of the past had come back to me. I feared that I would feel like this for the rest of my life—alone and trapped in a mind empty of memories. Were they buried somewhere I couldn't retrieve them, or had they simply evaporated?

Every day I'd wake up and hope that they had returned, only to be met with mounting anxiety and nagging reminders that I still didn't know anything I used to know. The sleepless nights were starting to take an emotional toll as well. The time I'd been watching TV, trying to learn while everyone else slept, was now spent crying. I could feel the depression creeping into the Swiss cheese of my brain and the rest of my body. And with the helplessness came hopelessness.

Joan said after all my surgeries I'd prided myself on being ahead of schedule for recovery, taking half the typical amount of time to heal and walk comfortably. That was not the case now. But what didn't compute was that my injury seemed relatively minor. I hadn't had brain surgery or just come out of a coma after fighting for my life; I'd simply fallen, hit my head, and lost consciousness. So what was taking me so long to rebound? I asked myself this question almost every waking hour, and I had quite a few of them these days.

. . .

Trying to battle the depression, I went into solutions mode and decided I was ready to visit my office in Tempe. Still, the thought of going to the place where I had built a company, the place that became the site of my downfall, was more than a little stressful. I was edgy, sweating, and my heart was pounding before Joan and I even left the house. On the twenty-minute drive over I had to fight to remain calm.

It's no big deal. I'll get through this.

Joan must have noticed my nervous fidgeting as I jiggled my heel up and down because she grabbed my hand. "It's going to be okay," she said calmly, looking me in the eye when we came to a stoplight. "I promise."

"Okay," I said, agreeable as always, trying not to alarm her. But inside I was still a jumble of nerves, happy to have her at my side.

We drove past Arizona State University and pulled up to my office complex, which was, as Joan described, a beautiful building that overlooked a lake, the ASU campus, and its football stadium.

We got into the elevator, and Joan hit the button for the ninth floor, explaining that we rented two adjacent offices there from Regus, an executive office company that catered to small business owners like us. Each company could rent its own fully furnished office and share several conference rooms, a kitchen, and office equipment.

Joan filled me in on who we would probably see at the reception desk. As she had predicted, the two office managers, Steve and Hollie, were at the front desk when we arrived. Both were aware of my condition, and Hollie had helped Joan get the accident report from the building management company.

After introducing themselves to me, they asked how I was feeling. "If there's anything we can do, all you need to do is ask," Hollie said.

Joan led me around the corner and unlocked my personal office, the larger of the two.

As I looked around, Joan pointed out what she thought I'd want to know. The first thing I noticed was the amazing view of the Tempe Town Lake and the magnificent rolling mountains in the distance, which I later learned were called the Four Peaks, McDowell, Camelback, and Superstition ranges.

I took my time, memorizing where everything was, and examined each picture and piece of artwork on and around the U-shaped reddish wood desk, which took up most of the room. The wall was hung with a document titled "Vision and Mission Statement," from my most recent company, Legendary Jets. Joan said she and the kids had gotten it framed for my birthday along with two pictures of me on my boat, *No Plane, No Gain,* which hung alongside it. Next to the giant wall clock and the flattened globe map were several photos: an autographed picture of news anchor Sam Donaldson, with the inscription "Scott, Smooth Flying, Thanks, Sam Donaldson"; a signed photo of astronaut Neil Armstrong standing in front of a World War II aircraft; and a shot of a medical team posing next to our Learjet 25, which had flown a heart in for the Mayo Clinic in Arizona's first transplant surgery on October 19, 2005. They gave our pilot a red heart-shaped pillow to commemorate the occasion.

My desk was stacked with papers and files, a fourteen-inch red and white wooden model of a Learjet 60, two of which I'd managed before the accident, a recent photo of Joan and me sitting on the boat, and a glass frame displaying back-to-back shots of Taylor and Grant. The room looked as though I'd left it moments ago because no one had touched anything in the month since I'd gone down to get my briefcase and muffin.

Looking out the window, I was awestruck by the magnificent view of the mountains. I was surprised at how high above the ground we were; our house was only a single story, and I'd never really stood at the hospital window and looked out.

At that point Anita came in, teary eyed, from the office next door and gave me a hug. Anita, who in her late fifties had short dark hair and a Boston accent, was very feisty and reminded me of Judge Judy. Grateful that she'd been helping Joan with our finances lately, I was happy to see her. I had spoken to her only once or twice, but Joan talked to her on a regular basis.

"It is good to finally meet you again, Anita," I said, repeating the standard line I had come up with for people I used to know.

With that, Anita left the two of us alone and Joan shut the door. I sat in my gray tweed swivel chair, and Joan took a seat across the desk from me. After I'd captured a mental picture of the office, vowing never to forget it again, I asked Joan a very important question: "Okay, now what do I do here?"

I was expecting an easy answer, even though I really had no idea what was involved in the day-to-day operations of running a business. Judging by the stacks of papers and all the file cabinets here, it looked as if I'd been busy doing a lot of important work, but I still didn't understand what I actually *did* at this desk every day. Tony Soprano didn't do anything but eat or talk on the phone at his desk, and neither did other TV characters. I also didn't notice any customers coming in to any of the other offices on my floor.

Did I sit and wait for the phone to ring? Did I initiate business contacts? Did I use the computer all day? Or was it more complicated than that?

My question must have surprised Joan because her jaw dropped and she looked as if she was about to cry. I could tell she was trying to keep her composure, but I thought she was going to lose it. She managed to answer my questions, though, her voice cracking with emotion as she tried to explain for the first time exactly what our company did.

We still managed corporate jets for clients, she said, after selling our jet charter business company about a year ago. We'd been developing Legendary Jets for the past eight months, designing all new marketing materials, a website, and new contracts. We'd also launched a new product called a Go Jet card, which offered blocks of time in charter aircraft in twenty-five-, fifty- or one-hundred-hour increments. Joan was usually good at simplifying concepts for me, but this was all over my head. I think I heard every fifth word. She seemed to know everything about my business, and I didn't even know where I was without her help.

At that point I needed a bathroom break, so Joan walked me back to the reception area, toward the elevator, and turned left at the corner. I had my hands on the lever to the men's room door when it struck me.

This is where I got hurt. This is where my life changed.

Even though I was on the ninth floor, and I knew I'd actually fallen in the first-floor restroom, it didn't make any difference because the layout was the same. As I walked inside, I examined the tile floor and tried to imagine how many steps I must have taken and where I had hit my head. I'd never really thought much about the incident until that moment, but to be honest, it wasn't anywhere near as stressful as I'd thought. In fact, it felt good to have conquered another fear—that I would reinjure myself or die if I went into a men's room in that building again.

After I had done my business, Joan was waiting for me in the hallway. I'm not sure if she'd anticipated I would have such an epiphany, but she must have guessed what I'd been thinking in there. "Are you okay?" she asked, assessing my face and body language and realizing, perhaps, that I'd done better than she'd expected.

I told her I was fine, and we returned to the office to order lunch from the café downstairs. Joan, who said I always used to buy lunch for Anita and Robyn, got the usual turkey sandwich for herself and Anita and a crab salad and cup of tomato basil soup for me.

While Joan was picking up the food, a man popped in, sat down in the seat across from me, and started talking. He was a big guy, about my height and age, wore glasses, and was dressed in a shirt and tie. He seemed comfortable in that chair, as if he'd sat there many times.

"Where you been?" he asked. "I haven't seen you in a while."

"I've been sick," I said.

"Well, have you been following this market?" he said, immediately delving into the bank bailouts and tanking stock market.

It felt like the temperature had risen to one hundred fifty degrees as the sweat beaded on my forehead. This was another one of my fears, and it was facing me head-on. Even though I'd been watching the news, I didn't get two-thirds of what this guy was saying about his 401(k), which I thought was a room number, and the effect of the economy on his profit sharing. On top of that, I didn't even know this man's name.

I know I must have appeared stupid, but I didn't know enough to respond, so I just listened, feeling immensely unprepared for this situation. Thankfully, Joan appeared in the doorway with our lunch to rescue me.

She must have sensed what was going on because she deftly distracted the man and got him to leave. "Scott hasn't been feeling that well," she said. "It's probably best if he just rests. We're going to eat our lunch and then head home."

As soon as Joan shut the door, I broke into tears. I told her that he'd just started talking to me and I didn't know what to do.

"I'm sure you did great," Joan said, hugging me. "I bet you he didn't even know."

But I knew better. I had failed miserably, or at least I felt like I had. Joan had left me alone in my office for only ten minutes, and I'd already had to deal with the embarrassment of not knowing someone.

During lunch I realized that I should expect to run into this dilemma countless times, and I needed to find a better way to deal with it because I had no desire for a repeat performance. I couldn't hide my memory loss, but I could avoid getting into a similar position by choosing who to talk to. Next time, I decided, maybe I could say I had to make a phone call or something and head off a disastrous one-way conversation like this one.

After we finished our meal Joan and I headed home. My headache had become unbearable, and I just wanted to hide, never to be seen in public again.

. . .

As time went on, I watched Joan get increasingly upset that my accident had robbed us of a lifetime of shared memories. I could hear the anger in her voice and hated to see my strong wife crying more and more.

"What if they don't come back?" she asked. "Someone's got to pay for this. You can't just slip and fall in an office building because of someone's negligence and have nobody do anything about it. We've incurred a lot of medical bills, and no one seems to care."

In turn, her anger started eating at me and made me feel tremendous guilt, which pulled me even deeper into depression.

How could I have let this happen to my family? I not only can't take care of them, I don't even know how, and the longer this goes on, the further into the red we'll go. How long can this continue?

Caught in a vicious cycle of self-recrimination, I felt sure that the old Scott would have been able to figure a way out of this; the new Scott didn't have a clue. The only thing I could do was ask how much money we had in the bank and how long it would last.

"Well, if the boat would sell, everything would be fine," Joan said. But she had another plan in mind as well. Looking for someone to take responsibility for my condition and pay our mounting medical costs, Joan called John Lohr, an attorney who had represented us and our companies for several years. After hearing that I was no longer able to run Legendary Jets as a result of my fall, John suggested that we contact the building owners to see if they carried liability insurance that would at least cover the medical costs. Suing for emotional and other damages could come later.

"If there's anything I can do to help, we do have an attorney here who specializes in personal injury," he said.

Joan and I discussed whether we should hire John's firm to explore suing the owners for negligence, but, as she tried to explain some of the legal terms involved—liability, Med-Pay, and litigation—she totally lost me, so I told her I'd leave it up to her on how to proceed. "I'll support you every step of the way," I said.

She hired John's firm, which contacted the owners, and after we got no relief there, we filed a lawsuit, which was settled in 2010. In the meantime, the doctors' bills continued to mount—along with Joan's anger, my depression, and our mutual frustration.

11

WITH ALL THESE FINANCIAL WORRIES, I really needed something more positive to think about. What better topic, I decided, than the woman who was caring for me on a daily basis and our twenty-seven-year relationship?

I'd been observing Joan closely for six weeks now, watching her every move and going through our respective things, trying to learn more about her—and us. I looked more closely through the bathroom cabinets one day and was amazed at how many bottles of lotion and nail polish, makeup jars, and hair curlers she had, versus my simple collection of deodorant, cologne, toothbrush, and razors. Her odds and ends took up three times as many drawers and twice as many cabinets as mine.

My conclusion: "Women need a lot more stuff," I told her.

Joan was no longer a stranger I was living with; she'd become my central life force and the key to rediscovering my lost past. She not only had taken over my former role as head of the household, she was also spearheading the medical quest to diagnose and treat my memory loss. Neither of us liked this switch because she wasn't comfortable with being in charge, but we both knew she was the most qualified person for the job.

We'd shared enough decisions in the past that I had to trust that she'd make the right ones now. Besides, she had become my best friend, the only person to whom I felt comfortable expressing my emotions and exposing my true self. I also sensed that she needed to feel that I needed her and felt an emotional attachment to her.

"It's okay to cry," she'd say, almost like a therapist, helping me to make sense of my feelings and to accept that they were understandable given my situation. "You can tell me anything. What are you going through?"

These conversations helped me feel that I wasn't crazy after all, and even though I felt weak and vulnerable, they gave me the strength to move forward. She never said this, but she seemed so appreciative of how open I was that I suspected—and she later confirmed—that I hadn't acted like this before.

I'm not sure whether it was because of or in spite of all this, but my feelings for her had slowly evolved from trust and need into something stronger. I wanted to spend more romantic time alone with her than before, without any agenda or lessons about who or what I was supposed to know. I felt different inside when I saw her and happier when I was with her.

I'd been chalking up the traits I most liked and admired about her. She was so loving and had so much integrity. It cracked me up when she laughed and no sound came out or when she made faces, reenacting lines from our favorite movies. She was skilled at finding the right medical care for me. Part mother and part wife to me, she also seemed to be a good mother to the kids; they surely seemed to love her. She enabled me to grow without making me feel self-conscious, sensing when to help and when to let me struggle. She had the bluest eyes and the warmest smile. I loved watching her put on creams at night to protect her skin, and even though she was a neat freak, I too liked order in the house. Her only shortcoming was that she could get a little disorganized and unfocused at times, trying to do too many things at once. But that was nothing in the overall scheme of things. She meant everything to me.

Lately I'd been reacquainting myself with my previous musical taste via my iPod, which contained some three hundred songs ranging from country to classic rock and alternative. One of them, titled "Back Where I Come From" by country singer Kenny Chesney, made me feel lost, yet I almost needed to hear the bittersweet lyrics, which hit me in the gut. The singer reminisces about growing up, raising hell with his friends in Tennessee, and what those relationships still mean to him. Well, I'd lost all those memories, so I would never know how those friendships formed who I was, let alone be able to experience them again.

Joan shared the songs that had meant the most to us since college, and "Faithfully" by Journey best described the essence of our relationship. Joan had been through so much with me since the accident. She'd continued to stick by me, and I'd promised her I would be faithful to her for the rest of our lives.

One of my other favorites was titled "Take Me There," by country singer Rascal Flatts, about a man who wants to learn about the woman he's in love with. This one didn't hurt me to listen to; it inspired me because I related so much to the lyrics: "I want to know everything about you, I want to go down every road you've been."

This was exactly how I felt about Joan. If I wanted to put together the jigsaw puzzle of my missing life, it made sense to start with her, one of its crucial corners. Who was she? Where did she grow up? What was our first kiss like?

One evening when we were alone, Joan curled up next to me, her head leaning back on a pillow so she could look into my eyes, her legs diagonally across mine. It seemed like the perfect moment. My memory wasn't returning, so it was time to start pulling the other edge pieces from the puzzle box and putting the frame together.

"So tell me about the time we first met," I said.

Joan's voice was bubbly and her face was glowing as she started at the beginning. We met, she said, at my college roommate Jeff's party in South Holland, Illinois, on a warm night in 1981, the summer after my freshman year. The party fell on my birthday, July 25, and although I'd been set up on a blind date with a friend of Jeff's girlfriend, Barb, the girl was a no-show. Coincidentally, Joan had also been set up on a blind date by Barb, a friend from high school in Tinley Park. When Joan showed up, I thought she was my date for the evening and was disappointed to learn that this cute girl was someone else's setup. After being stood up, I spent the evening talking to the same friends I saw every day.

At this point Joan stopped the story to tell me coyly, "If you'd told me it was your birthday, I would have given you a birthday kiss, and maybe my blind date would never have happened."

We didn't meet again, she said, until the fall, after a long day of football training camp. With my head freshly shaved as a bonding exercise among the offensive linemen, I joined a few teammates at a fraternity party on campus. Joan ran up and started talking to me, and we chatted for a good hour or so, getting to know each other as she stood a couple stairs up from where I was standing so we were at eye level with each other. As part of the starting lineup, I was exhausted from doing two-a-days, or practicing twice a day, but from what I told her, I was focused and driven to succeed.

That fall, Joan said, I took a class on the fundamentals of track—an elective for my physical education major—in which one of her gymnastic teammates was a student teacher. Apparently I asked out this girl, a cute redhead from New York, who was extremely flexible. Joan emphasized this point suggestively, but the remark went over my head.

"You were my second choice then," I said, prompting Joan to chuckle sardonically.

"No," she said, "she turned you down."

Humbly speaking, I couldn't imagine this happening, and although I had no way to prove her wrong, I suspected that Joan had thrown in that detail to tease me.

"It's a good thing she said no because I never would have dated you if you'd gone out with her," she added quite seriously.

"Why?"

"Because we had a rule that we didn't date a guy who had seen others on the team," she said, explaining that with their group of only ten women, it would have been too awkward otherwise. That didn't make sense to me—one date with a guy didn't mean he was your boyfriend—but I didn't question it further.

She said Jeff and I used to hang out by the athletic training room at the football stadium, where the football and gymnastics teams both got taped up before practice and iced afterward. Jeff was a fellow offensive lineman, and after we finished our training, we stood in the hallway, watching the girls practice on the uneven parallel bars, bounce on the trampoline, or flip around on the tumbling floor. Our cover story was that we wanted to see what kind of talent they had, but really we just enjoyed watching the girls bend and contort themselves in skintight leotards. The girls were no better; they liked watching us wander around in nothing but shorts.

Barb had talked to Jeff about setting Joan and me up on a double date, but Jeff never discussed this with me. So Joan, thinking this was a done deal, came up to me after practice one day and asked, "So when are we going bowling?"

Knowing nothing about the setup, I thought this sounded like a good idea. "Whenever you want to," I said.

This whole story made for quite an amusing discovery: Joan, my wife of almost twenty-five years, had pursued me, not the other way around. I had to rub it in. "So *you* asked *me* out?"

After we both laughed over this, I also had to ask her about that stupid rule again. "You really wouldn't have gone out with me if your teammate had said yes to a date with me?"

"Nope," she said adamantly.

As it turned out, we did go bowling on our first date. I know this doesn't sound very romantic, but we were poor college students. "I let you win so you wouldn't feel bad," she said. "Then when we went and ordered a pizza and sodas, you were short a dollar, and I had to lend it to you to cover the bill. Then I asked for it back on our next date."

"Nothing has changed in twenty-five years, I guess," I joked.

Afterward, Joan said, we went back to my dorm room and talked into the wee hours. We had an away game later that day and had to leave by 5:00 A.M., but I never mentioned the early departure time to Joan or she wouldn't have kept me up that late. We broke it up around 3:00 A.M., and I walked her three-quarters of a mile back to her dorm room.

When we got to her door, she said, I asked if I could kiss her good-night. After dating other friends of Jeff's who had been total jerks, she said, she was amazed at how a big guy like me could be such a gentleman. "I knew right then that you were the guy for me," she told me.

I could see the passion in her eyes and hear it in her voice as she said this, almost as if she was experiencing that moment again for the first time along with me. It took my breath away. Even though I knew intellectually that I'd spent the better part of my life with this woman, I'd really only met her six weeks ago, and I was starting to fall in love with her all over again. Before, I'd been yearning to feel these emotions, and now I actually was. I knew she wouldn't still be here with me if we hadn't formed this strong bond together. It wasn't the kind of tie you could maintain by yourself.

Ready for the next topic, I asked, "How involved were you in my football career at NIU?"

I knew she'd have a different perspective from my teammates about this, and I figured I'd eventually get their perspective. But more than that, I wanted to test her—and myself—to gauge how close we'd been before my accident by exploring how much she knew.

"Very," she said.

As Joan recounted the next series of stories, I could see that she'd been there with me—and for me—every step of the way. She even knew what I'd been like before we met. I got out all my crazy stuff in my freshman year, she said—lots of dating and drunken parties, getting in fights and waking up in cornfields—so I was ready to have a girlfriend by my sophomore year.

Joan said she came to many of my practices, we saw each other almost every night, and she traveled to almost all my away games in Michigan, Kansas, Wisconsin, and Ohio with my mom and dad, who never missed a single game my entire career. Hearing this, I figured Joan must have loved me from the start. Why else would a girl travel that far to see some guy play football every weekend when she could be enjoying parties at school?

Although football was important to me, Joan said it wasn't the sole focus of my life. Though I was talented and knew I was good enough to play in the NFL, always doing my best job on the field, at the end of the day I left it behind. Unlike some of my teammates, I didn't stay after practice to watch film, work out some more, or punish myself if we'd lost a game. I simply grew more determined to do better next time. I also grew to fear and respect my coach, Bill Mallory, who instilled in me the discipline that I applied personally and professionally for the rest of my life.

During my senior year, she said, we were both excited when I was elected one of three team cocaptains. After winning the conference championship in 1983, our team at NIU went on to play in the California Bowl in Fresno, the most important game of my career up to that point. The last bowl game for NIU had been the Mineral Water Bowl in 1965, when they lost to North Dakota, and it had been twenty years since NIU had won a bowl game.

"You were so proud of your team, and you knew that you were going to win this game and make our university proud," she said.

Joan came out with my parents to be by my side the week of the game. If I played well and we won, she said, it could make a big difference during the upcoming NFL draft. The better I did, the earlier round I could be drafted to a good team, with more money in my contract.

"You played really well, and the team won the game," she said.

After the game I was honored to be recognized as a member of the Mid-American Conference First Team All-Conference, which included the best players in my division—a sort of who's who on the twelve teams—and won an honorable mention for the All-American Team, its nationwide counterpart.

As happy as I was to hear all these stories, I was starting to shut down with the frustration of not remembering any of what she was describing. Joan, who had become well versed in recognizing this pattern, began summarizing the next phase of my football career.

When the season ended, she said, scouts from about fifteen NFL teams came to the university to watch me work out and had me do fitness tests to see how well I performed. Joan said she was very involved in helping me pick my agent, Jack Wirth, but she couldn't remember much about him except that he was from the Chicago area. She said we both liked him, and he seemed to know his stuff.

When Joan told me about the NFL Combine Camp in New Orleans, I knew exactly what she was talking about because I'd been watching live footage of the current camp on HBO and the NFL Network, trying to learn more about my past. As a result I could vividly picture my life as a college draftee.

The scouts had predicted that I would be drafted as early as the second round and no later than the fifth. (I later learned that the sports trade magazines had also published mock draft choices, which predicted that I would go in the fourth round to the Minnesota Vikings.) These were exciting times for us as a couple.

"When I walked around campus, your teammates called me Mrs. Bolzan," she said proudly.

The draft was scheduled for the first week of May 1984, the same month we were getting married, which meant that we had to prepare for our wedding, on the 26th in Joan's family church in Tinley Park, Illinois, amid complete uncertainty about where we were going to land later that summer.

The two of us spent draft day sitting in our apartment and waiting for the phone to ring.

"Why just the two of us?" I asked.

"Because that's the way you wanted it," she said. "You wanted to share this experience with just us."

It turned out to be a heartbreaking day, she said. We watched each round go by on TV, disappointed when the phone rang and it was friends or family calling. With each round I grew increasingly distressed, especially after we got into the sixth and seventh rounds. Finally, in the ninth round, after twelve hours of pacing around and passing the time any way we could, the New England Patriots' front office called to say I'd been selected and that I would be contacted the next day with details.

"We were devastated. All your hard work, everything we were told from the scouts, agents, and coaches. New England was probably the worst place you could have gone," she said, explaining that several All-Pros, or veteran elite players, who were supposed to retire had chosen not to, and the team had selected two other offensive linemen in the seventh and eleventh rounds, which meant that my chances of actually making the team after training camp were greatly diminished.

Joan sounded like she felt sorry for me as she recounted the events of that day, but all I could think was how fortunate and proud I was to have been chosen at all.

Wow, I was drafted. That's pretty good!

This was the dream of every kid who had ever played football, and looking back now, I was sad for the old Scott, who felt angry rather than lucky, regardless of the false expectations he'd been given. Of course I had no idea how I felt at the time, but I figured that I must've carried that anger into the training camp and had probably hurt my own chances for success there.

After ninety minutes of listening to Joan's stories, I'd reached my saturation level, so she summed up by saying that I was disappointed after getting cut from the Patriots' final roster in the last week of the preseason games, just as we'd predicted. I went on to play for a year in the USFL with the Memphis Showboats, she said, eventually returning to the NFL to play for the Cleveland Browns and ending my career with a foot injury in 1986.

This evening of stories only strengthened my feelings for Joan, knowing that she'd done more than anyone could or should have done for me. I was so thankful and grateful to have her in my life, someone who knew me better than I knew myself and could share these precious moments—both hers and mine—about my lost past. I had to wonder how all this anger and sadness she described had affected our life together, but the bottom line was this: this woman had stuck with me and supported me through the best and worst of times, and by the sound of it, she was in for the long haul even now. If anything, my injury had only sealed my love for her forever.

12

THE NEWS ABOUT GRANT'S DRUG PROBLEM had been weighing on me for some time, and the recent conversation with Joan about how we'd met had helped take my mind off it. But even though I was in information-seeking mode, there were some things I didn't want to hear about, such as the ugly lawsuit a business partner had filed against us.

"That was a very difficult time in our life because of the tremendous amount of stress," Joan said. "It ended up costing us our home and forcing us to file personal bankruptcy."

I stopped her right there, not knowing—or caring—that she hadn't even gotten to the worst part of her story, which I wouldn't learn for months. "I'm not in the mood to hear this negative talk," I said. "At this point in my recovery, I just have to concentrate on the good things."

I didn't want to spend any longer than I had to in the dark recesses of my mind. I also figured that I had plenty of time to hear about the bad parts of our life—later. I knew Joan was just trying to underscore that we'd always been there for each other, no matter what, and always would be. Still, I needed to recharge my batteries with some positive talk.

So Joan switched gears and told me about how I, as a Leo, the king of the jungle, had always tried to protect her and the kids from pain or harm. After losing Taryn, for example, she said we were at the mall when I saw a woman with a newborn approaching and guided Joan in another direction before she could see them pass by.

But she also said that now I was nowhere near as detail oriented as I used to be—remembering the smallest and most obscure facts after reading an article. I not only remembered the color of her shirt on our first date but the designer icon on it too. Now, it seemed, I didn't even notice these little details. I often had to re-read articles repeatedly before I could understand them, and even then I wouldn't remember much.

These stark changes in my observational skills gradually prompted Joan to realize that I couldn't have known who she was in the emergency room right after my accident. "It just dawned on me that you really didn't recognize me," she said.

"No, I didn't," I confessed.

"Why didn't you tell me?"

"I don't know," I said. "I didn't want to hurt anybody."

My secret was finally exposed, but I figured it was better to be honest about it. How else was she going to know what I needed help learning again? Still, we found it curious that the "protector" aspect of my personality had remained intact.

. . .

Following up on Dr. Fife's referral, Joan and I went for an exam with Dr. Heather Caples, a neuropsychologist at St. Joseph's, leaving the house an hour early because I was panicking about being late. Joan said I'd always hated to be late, calling it "Mallory time" after my college football coach, who had considered us late if we weren't fifteen minutes early. It helped me to think that my worries had some sort of rationale from the past behind them.

Caples and a female postdoctoral resident came to meet us in the lobby, and they took us back to a room for the testing. When Caples asked about my accident and the battery of tests I'd undergone since, Joan supplied the answers while I tried to calm myself. But as she started to characterize all the important long-term memories I'd forgotten, I could feel the stress rising in my body, like a fountain ready to spew forth with plumes of water.

When she got to Taryn's stillbirth, I began to cry uncontrollably.

"I'm so sorry for your loss, that must have been very painful," she said, pausing and apparently waiting for my response.

But the words would not come out. I could hear her speaking, but I was paralyzed by my emotions. It felt like an hour that I sat there weeping, even though it was probably only a couple of minutes. After gathering my composure, I told her that it hurt to be unable to recall something so devastating for both of us.

"How can I not remember losing our first-born child?" I asked. Once I choked this out, I was able to proceed with the testing.

After showing me patterns of triangles, circles, and squares, Dr. Caples asked me to draw them from memory in their same positions, had me do something else, then asked me to draw the pattern again. The second time I was able to add some new shapes I hadn't recalled the first time, but in this case my memory wasn't the problem; my hands weren't cooperating as I tried to replicate the images.

In the Boston Naming Test, I looked at line drawings of objects or animals, and had to name each one. I got forty out of sixty correct, failing to recognize some familiar objects such as a rhino or a hammock.

She read me a series of commonly used words, and asked me to recite back as many as I could remember. That was easy enough, and my confidence increased. But I had a tougher time when she talked about something else and then asked me to repeat the words. I felt my confidence sink when I couldn't remember them all. I hated feeling stupid. Caples also had me do a finger-tapping test to gauge my motor skills.

After about two hours of talking and testing, my head was killing me, so I had to take some pain medication, which I was still using every four to six hours. Then, while Dr. Caples talked with Joan in the next room, the resident drilled me with a series of question-and-answer tests for the next forty minutes. Her thick

Indian accent made it difficult for me to understand her, so I often had to ask her to repeat things.

At the end of our session, Caples gave us some impressions about my condition, but said she needed to further review the test results before reaching any final conclusions. She said the tests showed that I was below average in my ability to learn and remember new information, I had some difficulty paying attention, and I also had some problems recognizing everyday objects. She noted that my headache and use of painkillers during the testing could account for some of my attention issues and slower speed in answering. The physical and emotional symptoms I'd been experiencing since my fall—such as the headaches, fatigue, and anxiety—were typical after suffering a concussion, she said, but my severe long-term memory loss and failure to recognize objects were not.

"What you're describing doesn't fit your accident," she said, explaining that such extensive amnesia was usually the result of more serious brain damage than my MRI and other medical tests indicated.

Dr. Caples brought up what Joan had told her earlier, that I'd gone through a bout of grief over Taryn's death thirteen years after the fact, when we were going through our financial upheaval. Caples said that delay could suggest a tendency to repress or suppress emotional trauma. Because I had no history of cognitive problems and because severe retrograde amnesia like mine had been known to occur in cases of an underlying source of psychological distress, she said, I could be suffering from a dissociative disorder, the result of a traumatic incident that had caused my brain to block out my entire past.

"I recommend that you see a psychiatrist or psychologist to rule this out," she said.

Joan said she found it hard to believe that I'd had any emotional trauma or stressors serious enough to cause such a mental block, emphasizing how strong I'd always been and how I'd always dealt so well with adversity. She said we were both quite used to the ups and downs of working in sales and in the business world for so many years.

Caples brought up the ugly lawsuit filed by our former business partner, which Joan had mentioned, asking if it had been stressful for me.

"No," Joan said. "He knew that was just a normal part of our business." She tried to explain that when you're dealing with a business that has millions of dollars in assets and could cause big losses to investors, some of them were going to sue to try to recoup their loss. But the old Scott always felt that taking that gamble was worth the risk. In fact, she said, those high stakes seemed to help me maintain my competitive edge.

This whole conversation was troubling to me, mostly because I could tell how much Caple's comments seemed to upset Joan, whose opinion as a medical professional I trusted. I was also scared of the unknown.

Traumatic incident? What traumatic incident? The only ones I know about are Taryn and the accident itself.

I left the doctor's office even more confused than when I'd gone in, and still didn't understand what dissociative disorder meant, so Joan tried to help me make sense of what Caples was saying: perhaps this was my brain's way of dealing with some tragedy that I'd never really processed.

But what I heard was that, basically, I was a nutcase. "Do you think I'm crazy?" I asked.

"Absolutely not," she said, adding that she completely disagreed with Caples' theory that I might have a dissociative disorder, but conceded we should see a psychologist to rule that out.

Joan had already made an appointment with Claire Kurtz, a clinical psychologist in Mesa, when my memory hadn't returned within thirty days. Joan's disagreement with Dr. Caples' assessment was reassuring, but just to make sure, I called my mom to ask if I'd experienced any trauma that we should know about. I'd seen

TV news coverage of the sexual crimes to young boys by Catholic priests, and wondered if I could have been a victim.

"Did anything happen when I was a child? Was I molested?" I asked, explaining the doctor's comments. "Tell me now."

Since my parents' visit, we had been talking more often by phone and getting closer with each call. So, when my mom insisted there had been no such trauma, I briefly wondered if she and Joan were telling me the truth, but just as quickly decided that I trusted them both and saw no reason for them to lie to me. As I began to research dissociative disorder on the web, I saw other indicators for the disorder such as depression, substance abuse, alcoholism, or a history of mental problems, and I felt better knowing that I'd shown none of these before my accident.

· · ·

Dr. Caples emailed me her report on the morning of January 29, but it didn't include any new conclusions. She confirmed her earlier suggestion to see a psychologist or psychiatrist to further evaluate me and determine whether I had a dissociative condition, and also to help me cope better with the "significant stress" I'd been experiencing. She also recommended getting speech therapy to develop strategies to help manage my daily activities, and that I continue to have help managing our finances.

· · ·

As it happened, my appointment with Kurtz, the psychologist, was set for that same day.

We spent two hours discussing my accident, my relationship with Joan, Caples's neuropsych report, and Taryn. Joan did most of the talking about how our baby's death had affected us twenty-one years ago while I cried and had to explain why once again.

Kurtz was easy to talk to and seemed interested in helping us both. I opened up to her about the darkness and isolation, my constant questioning who I was, and my uncertainty about how I would ever get through this.

At the end of the session Kurtz said she didn't think I had a dissociative disorder, but she also couldn't determine any other psychological cause for my amnesia. She suggested that I come back to discuss my depression some more, but I didn't want to.

If I still don't have my memory, what good is it to talk to someone about what I don't know? I need to learn the world first and how to express my thoughts before someone can help me understand my feelings. I keep hoping that one of these doctors will give me a magic answer, but all I seem to do in their offices is cry and I leave feeling more frustrated and hopeless than I did when I went in.

. . .

With our business stalled and no new revenues coming in, we had to let Anita go at the end of January. As Joan and I grew increasingly concerned about our finances—one area of crisis management that she didn't handle very well—I decided it was time to start getting rid of the belongings that seemed excessive given our current lifestyle. These same decisions were being made in households across America, but for different reasons.

I approached this issue pragmatically because it seemed like a black-and-white problem that even I, in my limited state, could solve. Take the Chrysler that was sitting in our driveway, for example. No one had even sat in it since I'd come home from the hospital; why not sell it to bring in some quick cash?

We'd already borrowed money from Joan's parents, who had loaned it to us willingly and without judgment, but this only added to my feelings of inadequacy.

Going through some things on my desk, I discovered an old Hewlett Packard 12c calculator gathering dust. Joan told me I'd had the device for more than twenty years, ever since I'd been a financial planner in Chicago. It was far more complicated than your typical calculator because it could figure out future values, payment schedules, and compound interest.

When I asked her to show me how it worked, she laughed. "The hell if I know how to use that thing," she said.

"How am I going to learn it if you don't know how?" I asked.

"Look it up on Google."

Sure enough, I was able to find an instruction manual for the calculator on the web and spent at least three hours learning the basics. But I figured the time was well spent because I was thinking about trading in some of our cars and wanted to calculate what payments we could afford as I replaced them with less expensive vehicles. At least that was the plan.

Needless to say, being able to figure out these technical instructions and apply them to my real-life struggles was the confidence booster I needed.

Ready for the next step in the plan, I started running through our inventory of cars—my BMW, Joan's Porsche, Taylor's Chevy Tahoe, our Ford 150 pickup truck, our Chrysler 300, and Grant's old Honda—and analyzing what was essential for our family.

"Do we really need all these cars?" I asked Joan.

"No, but we use our three here, and we use the Chrysler 300 to go back and forth to the boat in California so we don't put too many miles on our other three. We have the Ford pickup truck just in case we need it."

Asked if we owed money on any of them, Joan said just my BMW. "I guess we could sell the Chrysler 300 if we wanted," she offered reluctantly.

Taking that as permission, I started figuring out how to sell it. I knew I'd been in sales, so I must have had an aptitude for it. By Googling "How to sell a car," I found countless methods listed, but I was determined not to spend days focusing on this task.

I stopped after just two hours of research, partly because I didn't understand much of what I was reading to be able to differentiate between options such as going to auction on eBay, selling on Craigslist, or doing a private party sale. I asked Joan if we knew anyone in the car business, and she said yes, a friend and former minor league baseball player named Jeff Kipila, whom we'd known for five years. "Why don't you give him a call and see if he can help you?" Joan suggested.

After finding Jeff's number in my phone, I had to psych myself up to make the dreaded call to a stranger who knew me but not vice versa. Similar to what happened with Jerry Pinto, I tried to explain what had happened and he thought I was kidding. Not surprisingly, I had the same reaction and got upset. "I'm not joking. This is a real difficult time in my life, and I need some help," I said.

After he apologized, I told him I was interested in selling our 2009 Chrysler. Jeff suggested I call a friend of his who was the sales manager at a local Chrysler dealership. But first he told me to go online and pull up the value of the car on the Kelley Blue Book website, which I did by plugging in the options from the original sticker I'd kept in my file cabinet. I noticed that the car's value had dropped $5,000 from the $27,000 I'd paid for it six months earlier.

Jeff offered to give a heads-up to the Chrysler guy before I called over there.

"I would appreciate that," I said.

The next day I made an appointment to see Jeff's friend.

"What are you looking to get out of this?" he asked me.

Pulling out the Kelley price quote I'd printed out, I said, "It shows here $22,000 is the value, so $22,000."

He just laughed. "Well, that's not going to happen," he said.

"Why not?"

"Because we can buy them cheaper at auction and I have thirty on the lot now with a $4,000 rebate that wasn't available when you bought yours," he said. He explained how the economy had cut deeply into the automakers, Chrysler included, even with the bailout.

I'd been feeling pretty positive up until that point, but my good mood quickly plummeted. "So what will you give me for it?" I asked sheepishly.

"Probably around $19,000, but I'll need to look at it first," he said.

He took the car for a spin down the street, inspected the exterior for damage, then came back inside to confirm his initial quote.

"Well, let me go home and talk to my wife," I told him. "I'll give you a call later."

On my way home I called Jeff, and he said that $3,000 less than the blue book price was fair if I really wanted to sell it. When I told Joan, she wasn't happy about the prospect of losing $8,000 on our car after buying it only six months earlier, so now I found myself doing a sales job on Joan.

"Think of it this way," I said. "It's $19,000 in our pocket now, when we really don't need this car sitting out on the street. And it will relieve some of the stress you're feeling. So let's just do it."

"But it's probably the most comfortable car to go back and forth to California in," she said.

I told her what Jeff had said and noted that selling it to the dealer meant we didn't have to go through the hassle of posting it online or worrying about strangers coming over to the house.

Ultimately, she gave in. "Well, let's just go ahead and get rid of it then," she said.

When I walked out of the dealership later that day with a $19,000 check in my pocket, I felt pretty damn good, as if I'd earned my first paycheck. I'd done the research, I'd gotten people to help me, I'd put my trust in a friend, and it had paid off, relieving some of Joan's stress, not to mention my own. Rather than adding problems to the family, I'd actually helped ease one, and I'd done about 85 percent of the work on my own. "We're going to save on insurance too," I bragged to Joan.

I still didn't know much about the world, but I now knew how to sell a car. And that was a major accomplishment, my first since the accident. I couldn't have been more proud of myself—so proud that I wanted to do it some more.

With gas prices skyrocketing and Taylor driving so many miles between school, work, and cheerleading practice, we felt it would be wise to get her a more gas-efficient car than the Chevy Tahoe. Seeing that I wasn't driving much, I took over driving that vehicle, which seemed to fit me better than anyone else in the family, and traded in my 2007 BMW. After paying off the $45,000 I owed on it, I was able to buy Taylor a much cheaper BMW 328i with a good financing deal and walk away with $11,000 in cash. Taylor was still an inexperienced driver, and I wanted her to have a safe and reliable car that wouldn't break down rather than a cheaper clunker that could leave her in a vulnerable situation.

Although Joan loved her Porsche, she insisted that we sell it because we needed the money to live on. With her prodding, I traded her Boxster and our pickup truck and bought her a 2006 BMW 330i. This time I walked away with $24,000 in cash.

Now, as our bank balance dwindled, Joan was all for making these changes, and my salesmanship was improving with every transaction.

Baby steps, I told myself.

13

JOAN AND I HAD BEEN PLANNING for some time to go to Oceanside, California, where we kept our forty-six-foot Meridian 411 yacht. I'd only seen photos and was looking forward to checking it out in person, but the primary purpose of our trip was to spend some intimate time together. I knew we were going to make love for the first time since the accident; I just didn't know when or how it would happen. I also didn't know how good I was going to be at pleasing Joan, and I'm sure I had more performance anxiety than most men who were about to consummate their relationships with new girlfriends or wives.

While we were there, I figured I'd also try to meet up with a new broker. Coincidentally, we'd put the boat on the market a couple of weeks before my fall, thinking we weren't using it as much as we'd anticipated and should put the proceeds back into our business or buy some new real estate. The broker we'd contracted with for ninety days hadn't gotten a single bite, and now that we really needed the money, we wanted to find someone more aggressive.

But before I was ready to take the helm, I knew I was going to have to relearn how to operate the controls so I didn't damage one of our most prized possessions.

As our departure date approached, I pulled up the Meridian website, found a manual, and began to soak up as much information as I could with Joan beside me to help make sense of it. The manual included diagrams of the control panel, for example, and broke down the complicated series of steps required to start the boat. If I didn't follow the instructions properly, I was horrified to learn that I could flood—and ruin—the boat's expensive Cummins diesel engine.

"Does this sound familiar to you?" I asked Joan.

"Uh, no," she replied sarcastically.

Joan said we'd divvied up responsibilities, with her taking charge of the interior, such as the lighting, kitchen appliances, television and its satellite dish, toilets, and water tanks. I was responsible for starting the engine and operating the electrical panel and fuel systems. Because of my piloting experience, she said, I'd been able to learn to drive the boat in only a few lessons.

But that knowledge and experience didn't help me now, and because Joan couldn't answer my questions, she suggested I call our friend and broker Jim, who had handled our original purchase of the boat for $325,000 in 2007. "I'm sure he'll be willing to help you out," she said.

The thing was, I wanted to learn as much as I could by myself first, to avoid looking stupid by asking dumb questions. Determined, I spent hours reading the diagrams and dry technical instructions until they sort of made sense and in my mind I could envision using the controls.

I also studied the overall layout, including the staterooms, bathrooms, galley, and two cockpits, and read the documents I'd saved from the purchase, a detailed maintenance history, and an appraisal, which provided me with extensive pictures of the components.

Hours later, it sadly became clear that I still had a lot to learn before I could safely operate this boat and not break something that would cost a significant amount of money to fix. I realized I had to swallow my pride and let Joan call Jim for help.

Joan explained what had happened to me, then handed me the phone. Shocked to hear about the accident, Jim proceeded to reassure me that I'd been a fast learner and a good captain who took pride in the upkeep of my yacht. "You know, Scott, you're a real bright guy," he said. "Once you see everything, it's going to make sense to you."

When Jim offered to help guide me on our maiden voyage, I thanked him for his advice and his offer. "I'm sure I'm worrying a lot for no reason," I said.

Joan also told me about Davey, an eccentric old-timer with a great sense of humor, who owned the same model of boat as ours and even kept it at the same dock. He spent most of his days tinkering around on it and helping his neighbors, including us, with their boats. After hearing this, I knew he would be the one I'd go to for help if I needed it.

. . .

When we still co-owned jets for our business, Joan said, we sometimes flew into the Carlsbad municipal airport, which was a fifteen-minute drive from Oceanside Harbor. Other times we'd make the five-and-a-half-hour drive in our roomy Chrysler. But after selling it off, this time we drove out in my BMW. With Joan at my side, I didn't need to consult MapQuest.

To keep my mind off my three-month-old headache, I took the wheel and talked to Joan, who pulled discussion topics from a book she'd picked up titled *4,000 Questions for Getting to Know Anyone and Everyone.* Leading us through politics to sex, favorite foods, spirituality, and sports, the book opened up new avenues for Joan to tell me stories about our past.

We stopped in Yuma, Arizona, for a much-needed break, where Joan explained we'd made the trip so many times that we knew where to stop for a clean bathroom, coffee, and healthy food choices. She also mentioned that we had other routines when we visited the boat, and she started to giggle.

"What do you mean?" I asked.

"Well," she said, pausing, "we always go to King's Fish House for a great lunch, then we go to the boat, and after you hose off the outside and I organize our stuff on the inside, we—well, you know—celebrate."

She kept looking at me to see if I understood her inference, and I laughed nervously. She leaned over, kissed me on the cheek, and snuggled with me. Clearly, she had a plan, and I was happy to go with the flow, hoping I didn't ruin things with my headache pain. I wondered, though, was I going to remember how to make love? Would it be like driving the car, where my hands knew what to do? Was it going to come naturally or be a challenge? Was I going to be able to give her what she needed? I felt like a different person now, so would I make love differently too?

Time flew by, and we were soon approaching the harbor. Joan seemed to be testing me, seeing if I might figure out where to turn, but as usual, I had to keep asking her where to go next.

After lunch at King's—crab cakes for me and blackened ma-himahi for her—we pulled up and parked at the dock marked *T*, which was about twenty yards from the shore.

"Wow, nice. We have beachfront property, don't we?" I said, genuinely in awe.

"Only the best," Joan said. "The boat is our oceanfront floating Cali home that we always wanted."

We grabbed Joan's rolling suitcase, my duffel and black sports bags, and a grocery bag of clam chowder, pecan sandies, sodas, bread, and peanut butter, and headed down the ramp to the boat. Twenty-five vessels lined one side of the dock, and a several-story weatherworn condo complex, with balconies running across the beige stucco façade, hung with beach towels and flowerpots, over-looked the harbor on the other.

"We're near the end of the dock, past where the building ends, and have nothing but sea air in front of us," Joan said.

As we got closer to our slip, she said, "We like to start our week-end midweek—it is very quiet and *very* private," she added, with the same giggle and mischievous look as before. The other docks had boats on both sides, and she was right—our spot was more isolated.

Our sleek white yacht, with its tan leather upholstered seats and beige carpet, Formica-covered cabinets and furniture, and its entertainment center, complete with a color TV and CD and VHS/DVD players, was quite impressive. I felt proud to have been successful enough to buy it with cash.

Joan confirmed that the Meridian was the one purchase I'd found most rewarding, and because I enjoyed it so much I called it my sanctuary. It was nice to know that at least one thing in my life hadn't changed. I located the hose to rinse off the salty, corrosive residue from the ocean mist while Joan went inside to prepare the cabin.

Once I finished, I came in to get a drink of cold water from the fridge and was surprised to find Joan in the master stateroom, lying on her side in bed, with the navy blue cotton blanket and gold sheets pulled up to her chin. Why was she in bed while I was outside working?

Peeling the covers back a bit, she patted the bed next to her, and I started to get the idea. Sweaty and wet from hosing down the boat, I dropped my shorts and T-shirt to the floor. As I pulled back the covers, I was pleased to see she was wearing a fuchsia lace camisole and matching boy-cut panties I didn't recognize. She rolled over to show off her backside and the rest of her outfit.

"Well, that's new," I said.

"I have a lot of these outfits in different colors," Joan replied coyly.

"I think I want to see them all."

I climbed into the bed, which was elevated atop a chest of drawers, with two steps on either side of it, and modestly covered myself with the blanket. Joan had opened the porthole and top windows and turned on the fan, so a gentle breeze was blowing through. As she slid the covers down to expose my chest and cuddled up next to me, I started to feel warm all over.

"Are you still hot?"

"Yes."

Unsure if she was being literal, I watched her adjust the fan and felt her run her fingernails up my side, which caused goose bumps to erupt across my chest. Joan told me that I'd worked construction during summers in college, and when I came home to our apartment, which had no air conditioning, she would cool me down using this same method. It was working all right, but it wasn't the temperature that was making me sweat this time.

Joan sat up and faced me on the queen-size bed, and I took in her beauty as I ran my hands down her silky smooth arms. She'd brushed her hair and put on some of that Versace perfume I liked. Her skin had a rosy glow, and I wondered if she was as nervous as I was. When she smiled warmly, I felt safe.

"How's your head?"

"Shipshape," I replied with a chuckle.

She leaned over and kissed me softly, and I could feel some of my fear turn to arousal.

"With your insomnia, I'm sure you've watched a lot of Showtime in the middle of the night, and even though I don't look just like those girls, I don't think you'll be disappointed."

Joan was referring to the soft-core adult entertainment programming, featuring topless women having simulated sex with men, that ran in the wee hours. "Yeah, I've seen it," I said, "but—"

"—don't worry, I'll be gentle with you, my forty-six-year-old virgin. This might be fun. I'm also relying on your unaffected procedural memory to kick in somewhere."

And with that she kissed me again. I felt her breasts under the lace top brush against my chest. Imitating what I had seen on TV, I slipped the strap off her shoulder and began to caress her. Part of me wanted to lie back and be seduced; the other part wanted to show her that I was still a man who could take control. But as much as I wanted to feel manly, I still didn't know who I was, so mostly what I felt was confused. And, seeing that the ceiling was so low that my petite wife had only a foot of clearance above her head, it made sense to let her take the lead in this dance, at least for now.

"Are you okay?" she asked.

I looked down and said, "I think this means I'm more than okay."

Joan playfully controlled the experience, but she treated it—and me—with levity so it didn't feel like a lesson. I was still nervous but relieved that she was guiding me along as we moved forward. Better that than for me to take charge and get it wrong.

"I know you don't remember, but we've been down this path many times before, so just go with what feels good, and hopefully it will come back to you," she said. "And just remember, I love you."

Well, they say it's like riding a bike, but it isn't if you have a brain injury like mine. I was so preoccupied with doing things right, fumbling around and not knowing what order to do things in, I couldn't really enjoy the experience. Like many virgins, I just wanted to get the first time under my belt. The bonding feeling I'd heard about didn't come until afterward, when we were lying there, talking it over.

"See, I told you it would all come back," she said.

"Okay, if you say so," I said, wondering if she was just trying to make me feel better.

Even though it was hard for me to talk about this, I asked what I'd done right and what I could do differently, hoping to improve for the next time. I could tell Joan was trying to be encouraging, but I had to admit that she looked relaxed and happy, so I figured I must have done something right.

We'd been lying there, talking and snuggling for about forty-five minutes, when Joan started crying.

"What's wrong?" I asked.

"It's overwhelming," she said, explaining that she was happy *and* sad. Making love had been nice, but it was so different, so tense, and I'd seemed so scared.

Hearing her say that, I wondered if I'd ever be the same—if *we'd* ever be the same—and be able to please her as I had before. She kept saying we'd had a healthy sex life before; I was hoping that practice would help, and practice we did. I soon began to develop a repertoire of moves that seemed to work well, and I grew increasingly comfortable initiating them. By the fourth or fifth time, I felt better about my performance—and us—although, just as in the rest of my life, my confidence was still low and I took rejection personally. But once I'd finally tasted the proverbial apple, I realized how much I liked it, and started wanting it more—and more often.

. . .

We were still enjoying our alone time together when Grant called on Saturday and said he was struggling not to use drugs. "All I want to do is take all the money out of my account and go get high for a few days," he told Joan.

Every couple of weeks Grant had been calling with some crisis or other, not having enough food or money or needing a ride somewhere. His problems never seemed to end. In my view they were all the result of the life that he had *chosen* to lead, and I was tired of dealing with his problems when I had so many of my own. That said, I didn't dispute the way Joan was handling this—helping him whenever he asked—because she'd been dealing with him for much longer than I had.

She and I talked, and we agreed that Grant needed his mother there to help him through this. So while she talked to him on the phone, I busied myself booking her a flight online out of John Wayne Airport in Santa Ana. We decided that she would go home to deal with Grant while I met with the broker Sunday afternoon as arranged. I'd never been to this airport before, so I made sure to print out my MapQuest directions and also programmed them into my GPS system as backup because sometimes the two didn't match up. I was okay with Joan leaving to help Grant as long as I was consumed with the logistics of getting her home. It was only after she left, when I had time to reflect, that I felt lonely, thinking how nice it had been to get away and how abruptly we'd been interrupted.

How often will this happen in the future? Will he always come first?

But I was learning from Joan that you have to prioritize people's needs in life, which is what she told me she'd done as a triage nurse, and she seemed pretty good at it.

Later, around 9:00 that evening, Joan called to check in, but she was elusive about details. I let it go, figuring that she didn't want to worry me when there was nothing I could do from here. Surely she would fill me in soon enough.

After signing a contract with our new broker, I called to update Joan. "But more important," I said, "how's Grant doing?"

"He's having a real rough day," she said, adding that he was going to spend the night at our house.

I took myself out to dinner harborside, returning to the boat to watch some TV. The plan was to go to bed early and head home around 9:00 in the morning, but as I was watching the news, something very strange, unexpected, and exciting happened.

One by one, a series of memories from my early childhood in Chicago, about a dozen in all, flashed across my mind. These short glimpses revolved around a backyard barbecue, with a bunch of adults sitting in folding nylon and metal chairs next to a pool, a long skinny lawn, and a chain-link fence. None of the adults had faces, and although I couldn't visualize what they were wearing, it was clearly warm enough to sit outside. I could also see a big apple tree, a kid's bike at the bottom of seven steps that led up to a metal screen door, and a garage with power lines running across it.

Each flash lasted only thirty seconds over the course of maybe ninety minutes, but they were all so vivid and clear I was beside myself.

Yes! This is it! I have my memory back. It's all going to come back to me now.

Ecstatic, I wanted to call Joan right away, but I decided to save the bombshell until I could watch her face light up. It was too late to start driving home now, so I'd get a few hours sleep and leave around 3:00 A.M., which would get me home around 9:00 to start Joan's day off with a bang.

But things didn't work out as planned. When I pulled up to the garage around 8:45, Taylor was getting into her car to go to school—about an hour and fifteen minutes later than usual—and she looked upset.

"Why aren't you at school?" I asked.

"It's been a terrible morning. Talk to Mom."

"Tell me what's wrong," I said.

Taylor started crying. "I don't want talk about it, Dad."

"You are not leaving here upset. Now please tell me what's wrong."

"Grant relapsed again. He's using heroin now."

I almost passed out. I simply could not believe what my daughter had just told me. Heroin? Joan said she'd never seen heroin as a nurse, so I assumed that I'd never met someone who knew how to get it, let alone use it. Here we were the ones who were supposed to be older and smarter, and yet my son knew how to buy this street drug that dirty lowlifes used?

"Where are they now?"

"Mom took him to this free place in the ghetto for detox," Taylor said. "They just left."

I hugged Taylor until she stopped crying and told her she could stay home for the day. "No, I don't want to be here," she said. "I want to forget about this."

I called Joan but got no answer. She either wasn't picking up or her cell phone was dead; she often forgot to recharge the battery. My concern about not being able to reach her quickly turned into anger. I was so pissed that Joan hadn't told me about the severity of Grant's drug problems, I couldn't even see straight. I was convinced that she'd been trying to protect me from this problem, and this was not something I wanted or needed to be protected from.

Why is she doing this alone? Is this how life has always been?

I'd just gotten some of my memories back, and I wanted to share them with the most important person in my life, but she was too busy taking my son, the drug addict, to rehab.

An hour or so later Joan got home, and I confronted her as we sat at the kitchen table.

"Why didn't you tell me about this?"

"I didn't know Grant had been doing heroin until this morning," she insisted, explaining that he told her he'd pretended to be walking our dogs when he was actually walking to a drug dealer's house two blocks from ours. So much for the nice, clean neighborhood we thought we'd moved into. He'd apparently tried to self-detox before but had never made it past the second day, when he started getting sick.

So we've been lied to and deceived? For how long?

I still found it hard to believe that she hadn't known about this. "Is that the truth, or is that just what you're telling me now?" I demanded.

"No, really, I just found out this morning. I would never lie to you. I was going to tell you when you came home," Joan said. "I never thought you would come so early and run into Taylor. And why *are* you home so early? I wasn't expecting you until 11:00 or so."

These obviously weren't the best circumstances, but I couldn't wait any longer to tell her. "Well, I have good news," I said. "I got some memories back."

Joan started crying, she was so happy. "Oh, my God!" she exclaimed, her hands flying up to her cheeks. "Tell me, tell me! Your memories! They're going to come back. They're going to come back!"

After I recounted the snapshots of my past, I asked if anything sounded familiar. But because she didn't know much about my early childhood, she suggested I call my mother. I did, and she confirmed that the scenes I described were from our apartment in Calumet City, a small community south of Chicago, when I was five or six.

As Joan and I talked further, my initial burst of anger about Grant, which stemmed from the frustration of not being able to get enough information, began to subside. Once he got out of detox four days later, he went back to his apartment, where he picked up his life again and told us he was trying to stay sober. I felt torn, so happy about my memories coming back but still so devastated about Grant's increasingly serious drug problem. I only wished I could be as strong as Joan, who seemed to be able to put aside her obvious pain about our son's medical crisis in order to feel happy for her husband's medical breakthrough. It made me feel better when she confided that with her nurse's training, she typically went into a numb, emotionless action mode during times like this, saving the breakdown and mess of tears for later. Such was the case, she said, first when Taryn died, then when Grant suffered his head injury. But obviously I didn't remember any of that. My Joan of Arc was all I knew.

14

W HEN THE WEEK of Valentine's Day came, I didn't understand its emotional significance, so Taylor briefed me—at her mother's suggestion—and told me to get Joan a card. But she didn't give me a hint about buying flowers or a romantic gift for the woman I loved. It wasn't until I saw the dozen yellow roses arrive in a vase for Taylor that I felt like an unfit husband.

How could I not have gotten something for Joan, the woman who has done everything for me, shown me so much love, and has stood by me, taking care of me every step of the way since the accident?

My ignorance, once again, was hard to accept. After I'd told Joan several times how sorry I was, she finally said, "Please stop apologizing to me. We don't need to give each other gifts to know that we love each other."

These comments only reconfirmed how giving Joan was and that although she appreciated that we were still together after my injury, she was more interested in whether I'd fall in love with her again.

"I love you, and I love the fact that you love and care for me the way that you do," I said, embracing her and giving her a tender kiss. "Next time I'll know what to do."

. . .

As Joan's birthday on February 23 approached, she was becoming more tense, a stark contrast with how Taylor had acted near her special day. I could understand that celebrating birthdays was no longer as much fun for someone our age as it was a reminder that we were getting older, but I wondered if something else was going on.

"Have you noticed Mom is acting different, more stressed, lately?" I asked Taylor.

"She never really gets excited about her birthday because four days later is Taryn's birthday," Taylor explained.

"What do I usually do for Mom's birthday?"

"You've always tried to make it a special time for her," she said. "You've surprised her with a trip to San Francisco. Last year you took her to Chicago and surprised her with Bon Jovi concert tickets. You're always romantic, and you do your best even though she doesn't want to celebrate it."

I believed all that, but I knew this year had to be even tougher with the stress and depression she'd been experiencing since my accident.

"Maybe I need to discuss this with Mom before making any plans," I said.

But when I tried to talk to Joan, she wasn't very responsive. "I don't want anything for my birthday—maybe just go out for a nice dinner with the kids," she said.

She was clearly in no mood for festivities, especially knowing that she was going to have to endure the anniversary of the worst day of her life with a husband who couldn't genuinely share her grief. Sensing all this, I wanted to please her even more, so I agreed not to get her anything and that we would go to one of our favorite Japanese restaurants where they prepared the food tableside.

We all met up at Way Sushi & Teppanyaki around 5:30 and had a long table all to ourselves. Joan ordered her usual fillet and scallops, Taylor got the chicken and shrimp, and Grant and I ordered the scallops, which Joan said was my favorite.

The cook put on a show with fancy knife work, forming a volcano-shaped mound of onions that billowed white smoke after he poured oil into the center of it, then made a "choo-choo" sound as if it were a train's smokestack. That seemed to break the ice a bit. Our elevated stress level was compounded by having Grant there, in another one of his moods. But at least he wasn't acting out like he had on Taylor's birthday.

I tried to keep the general conversation light and upbeat, hoping to make Joan laugh.

"Don't worry, I'm still older than you," I said.

"Yeah, six months."

"Well, you still look better than women half your age."

I knew deep down that Joan was putting on a front just to get through the night. Taylor told the owners it was Joan's birthday, so they brought over a dessert plate of sliced pineapple wedges with a single lit candle and sang "Happy Birthday" in Japanese. I found this amusing, and Joan seemed to appreciate the gesture.

After dinner Joan's parents joined us back at the house for chocolate cake and coffee. By this time I felt comfortable enough to actually sing the birthday song. Taylor said we should put only one candle on the cake even if we had forty-six in the house.

When Joan had obviously had enough, her parents left, Grant went home, and Taylor went to Anthony's, leaving Joan and me alone. I told her I felt conflicted about not getting her a gift, but I'd honored her wishes just the same.

"I just don't feel very happy about myself these days, and I don't want to focus on me," she replied.

"I understand the way you feel, but I'm still trying to figure all of this out, and I want to make sure that I'm doing everything correctly."

"You don't need to feel disappointed," she said. "It's more about being with you and the family. That's a gift enough. Just get your memory back for my birthday."

"I'm trying."

"I can hit you in the back of your head again if you like," she joked. I immediately thought of the Three Stooges and figured she was going to be okay.

. . .

Among the voicemail messages that Joan had saved was one from a guy who said he hadn't heard from me in a while and wanted to have lunch.

"Did I know a guy named Mark Hyman?" I asked.

Joan reminded me that we'd been friends since we had offices next door to each other in 1993. She called him to explain what happened, and he and I had a short, awkward conversation. He said he'd wondered what happened to me because we'd planned to have a holiday lunch the day after my accident but I'd never called to confirm. He invited me to lunch again, but I wasn't ready yet to try having a conversation in public with someone outside my immediate family.

Joan had been encouraging me to rebuild my friendship with Mark and my old friend Jerry Pinto, saying it would be good for me to have someone other than her to vent to, someone who could give me a different perspective on the old Scott and could fill me in on private man things we used to talk about. "You don't tell your wives everything," she said.

Besides that, she said, who better to teach me how to be a man than another man?

These seemed like good reasons, and, frankly, I was intrigued to see whether my friends were anything like me. I knew that Jerry and Mark were both a decade older than me, and even on the phone they sounded far more boisterous and self-assured than I felt inside, which was meek and reserved, even though I didn't know what those words meant at the time. I didn't really know how to be myself with either one of them because I didn't know who that was, so all I could do was react to whatever they said or did.

Back in February Jerry had flown in for a brief trip to help Joan and me resolve our health insurance problems, but we didn't have more than an hour's conversation over dinner before he flew home with my signature. I'd been hoping to rebuild our friendship, but in spite of his promises to be available anytime to talk to me—"If you're upset, call me. If you can't sleep, call me"—I'd left multiple messages on both his cell and business numbers and he'd only called once since his trip.

"Sorry I haven't called back. I've just been busy. I'll give you a call tomorrow," Jerry said. But the call never came.

Initially I thought maybe he was just busy, as he said, but when the trend continued, I wondered what I'd said or done wrong, so I kept leaving messages. We finally did connect a few times later in the year, but they were brief conversations, with dozens of messages from me in between.

Mark, on the other hand, had been checking in with me every couple of weeks, and when I was ready, we made plans to meet for lunch at Chompie's, a Jewish deli in Scottsdale.

When Joan dropped me off, I saw a man fitting her description standing outside the restaurant and pointing at me. He was balding on top with brown, graying hair. Wearing nylon workout shorts and a T-shirt, he was about five feet ten inches tall and two hundred and forty pounds. If they made a movie about my story, I'd want a younger version of actor Abe Vigoda to play Mark.

Without Joan there, I had to order for myself for the first time. I knew I didn't like bread, so I told Mark I was going to have the chopped liver on crackers because it looked good in the menu photo. Mark, who is Jewish, was surprised and even more surprised when I ended up liking my meal.

Joan had told me that I could trust Mark, but I also sensed that on my own. I felt it in my chest—what Joan called my "gut instinct"—that my feelings were safe with him and that he wouldn't judge me. So I told him I didn't remember playing football or who Joan was in the hospital, and I confided in him about my constant fears and anxieties.

"It must have been really traumatic for you to lose not only your memory but everything that you've worked for and everything that you planned for," Mark said. "To lose your goals in life, who you are as a person, must be very disturbing."

"It's been very hard," I said, pleased to hear someone other than Joan be so warm and understanding. "I'm having to re-create myself."

Mark had three kids, even younger than mine. He and I were able to talk seriously about issues such as our families, my wife, and his ex-wife, and yet still joke around, including who was going to pay for the check. He offered to pay, and I let him because I didn't know what else to do. Joan later told me that friends often alternate paying for lunch and that I could get the bill next time.

From that point on, Mark and I met for lunch at least once a month. Sometimes he joked that I had done myself some good with this accident because I'd never been one to talk about my feelings before. But when he told me I used to be confident and enterprising in business, he might as well have been talking about a total stranger.

. . .

Around this same time Joan was recruited by someone she'd met at a charity function and was hired as the director of a hospice foundation in Phoenix, where she was to be in charge of fund-raising and operations, starting in mid-March. She said this was something she'd always wanted to do, but more important, she needed to start earning an income and reduce our health insurance costs.

I wasn't thrilled with the idea of her going out in the real world while I stayed home alone. With Taylor at school and Grant in his own apartment, the prospect of being alone for eight to ten hours a day was frightening.

What will I do all day? How will I get by? Who will teach me what I need to know?

Joan had rarely left my side for nearly three months. She'd provided me with the knowledge to exist, shown me how to live and how to love. For her sake I tried to laugh it off.

"You're going to have fun not having to deal with me five days a week," I told her.

She laughed. "I would still rather be here with you than go back to work," she said. "You're easy."

Joan was concerned about the career change ahead of her, given that her past work experience was as a registered nurse and as a marketing director for our jet company. Working for me was the easiest job she'd ever had, she said. She didn't have to show up for work every day, and she got to sleep with the boss, who always took her on vacations. But I figured that Joan was going to do just fine. After all, she had a master's degree in leadership. I didn't know what that meant, but it sure sounded good.

When Joan was getting ready for work on her first day, March 15, she looked happy and relieved. I figured she was pleased to be returning to work with her own age group, not having to spend every minute with her three-month-old husband. I had to admit I couldn't blame her. This was her time to get away from me and to share her experiences and knowledge with others.

Still, it was tough to watch her walk out the door in her brown suit and pumps, her Louis Vuitton computer bag, and a small box of family photos and knickknacks for her office. I told myself it was time to share her with the rest of the world; this was also my opportunity to discover new ways to fend for myself. I wanted her there with me every second of the day, even if we got on each other's nerves sometimes, but I had to admit that I really needed her to leave.

I walked her to the garage door and held her longingly, as if I wouldn't see her for ten years. "Have a good day at work, and don't worry about me," I said. "I'll be fine here by myself."

She sighed. "I know you'll be fine," she said. "I'm just concerned about leaving you alone."

I watched her back out of the driveway and head down the street before I walked into the house and closed the garage door. After Taylor left for school, the house was all mine, and it was oh so quiet.

I walked around the house, looking for some magical sign telling me what to do next. The house had never seemed so big. No one was there to stop me from getting into trouble or to tell me how to do things. I knew Joan was only a phone call away, but I was determined not to bug her. She had enough going on in her world, getting used to a new job and dealing with new people.

When she called around 11:00 A.M. to check on me, I was happy to hear from her.

"I'm okay, but I miss you," I said.

"Maybe I shouldn't be working yet," she said, sounding worried. "Maybe I should be home, taking care of you."

"I'll be fine," I told her again. "I need to learn on my own, and you need to be with other people so you don't go crazy."

That seemed to calm her down. "What are you up to?" she asked.

I told her I'd been reading news articles on the computer and going through my desk, scanning through documents and brochures to learn more about what I used to do for a living. I also felt it was important to learn as much as I could about the changing economy, the automakers' bailout, and the ongoing controversy over CEOs' use of private planes. I was trying to determine whether I would ever be able to make money again by managing private jets amid all this negative media coverage.

We ended the conversation with the usual "I love you," and she said she'd be home around 6:00 because of the traffic.

As a way to show her that I had everything under control in her absence, I decided to make a nice dinner of chicken parmesan with steamed vegetables and garlic bread.

After a trip to the grocery story, my headache was getting bad, so I took some pain medicine and relaxed in my chair until Taylor got home. I must have dozed off because the next thing I remembered was Taylor opening the door around 3:00 P.M.

"I'm home. How was your day, Dad?" she asked in her usual teasing tone. "What did you do all day by yourself?"

"It was fine," I said. "I didn't hurt myself or get lost."

"Well, that's a good day then," she said, laughing.

I asked if she would be home for dinner, and she said no, she had to work. After I told her what I was making, she replied, "You better save me some leftovers."

I cleaned up around the house, opened the mail, and read the paper until 5:00 P.M., then started on dinner. I'd already made this meal once since my accident after Joan ordered it for dinner in Oceanside, and I felt comfortable trying it without the recipe. Cutting the chicken in thin slices, I pounded it with a meat hammer, soaked it in eggs, coated it with Italian bread crumbs, then sautéed it with olive oil, garlic, and pepper.

While that was cooking, I boiled water for the rotini noodles and cut fresh zucchini and carrots, which I steamed and sautéed. While the bread was in the oven, Joan popped in around 6:00, just as she'd planned.

I ran over to welcome her with a hug and kiss. "Wow, what smells so good in here?" she asked.

"I'm making one of your favorite dishes, chicken parm."

"Yum."

"How was your first day?"

"It was fun. I think I'm going to like this job," she said, barely missing a beat before asking about me. "So tell me what you did all day."

I told her I worked, went to the grocery store, got rid of a headache, and watched a lot of TV. "We're going to switch to Geico insurance because we can save up to 15 percent," I said, proudly quoting the little green lizard.

"Scott," she said, laughing, "we already have good insurance, and we don't need to switch."

I was a little upset, not understanding the humor in what I was saying. I was seriously trying to cut our costs. "Look, if we can save money, why not do it?"

"Why don't you concentrate on learning the important things, such as current events, past history, things like that?" she replied.

Now I really felt insulted. She was telling *me* what was important? *Everything* seemed important to me. "So, I guess the ShamWow! is out of the question then," I snapped sarcastically, hoping to convey my hurt feelings.

Joan looked at me as if she couldn't tell if I was kidding, but, well aware of my limitations, she softened her tone. "Not everything on television is a good choice," she said.

As we sat down to eat the meal, which turned out quite well, I might add, she told me about her job, her co-workers, and how much she still had to learn. Then it was my turn. I had a new job as well.

"It's difficult not having you here to keep me on track and show me what I need to know," I told her, explaining that it was lonely not having her to talk to whenever I wanted. "I found it very distracting."

I wasn't trying to upset her; I could see that she felt torn. I knew she wanted to take care of me, but she also felt it was important to go to work. I wanted to reinforce that I was all right with that and that I would benefit from struggling on my own.

In the coming days I developed a routine: I read the paper over breakfast, played with Mocha to exercise her, straightened up the house, took out the garbage, or did laundry. I tried to learn more about my previous life and current events by going through files and boxes, Googling issues of interest I'd seen in the newspaper, and watching the History Channel and Fox News. Mocha soon switched her allegiance from Joan to me, nudging me to be petted and napping at my feet in my office. I enjoyed the company.

It could have been because of my headaches, my healing brain, or the pain medication I was still taking, but my attention span didn't seem to be expanding. I found myself needing to shift tasks every fifteen minutes or so, frustrated when I didn't understand something or when I'd reached a saturation point in learning the issue at hand.

When I tried to read a brochure from my business, for example, I could understand what it said, but I had no context or experience to comprehend its purpose or to know when I would present the pamphlet during a sales pitch. Was it a compilation of marketing approaches that had worked in the past, I wondered, or had an attorney drawn it up for me? Such unanswerable questions could drive me so crazy I'd have to switch gears.

As increasingly self-sufficient as I was becoming, I didn't want Joan to feel that I didn't need her, because I did. The more she was away, the more I wanted to be around her. I so hated being without her that I would wait by my computer as dusk descended, watching the cars approaching on our closed-circuit security system. As soon as her car pulled onto our street, I opened the garage and greeted her with open arms.

Every day I felt the bond between us growing even stronger.

. . .

The nights, however, were still a battle for me, and that was starting to affect my days. As time progressed, my insomnia, which, I'd learned from reading, was a common side effect of brain injuries, grew more erratic. In the first few months after my accident, I'd consistently been getting no more than three hours of sleep—an hour or two after I went to bed around 11:00 and the rest in five- to twenty-minute catnaps during the day.

A few weeks after Joan went to work, I was miraculously able to lie down one night and sleep for eight hours straight. But the excitement diminished a couple of days later when my previous insomnia returned. Although it varied, it was typically three to seven nights before I could knock down eight hours again. There was no cause or pattern that I could discern to this new sleep syndrome, and it became a vicious cycle.

Oddly enough, I found it easier on my body and mind when the insomnia was consistent than when I had to adjust to these extreme variations. I became increasingly fatigued, and my body ached all the time. My primary care doctor, Teresa Lanier, told me this could either be a side effect of the pain medication or simply a result of my brain injury, but she cautioned me not to nap during the day because I'd never kick the insomnia that way.

But as hard as I tried to heed her advice, some days I just couldn't make it through the day without a thirty-minute nap. Dr. Lanier prescribed the sleep aid trazodone, but it made me groggy in the morning. Next I tried Flexeril, a muscle relaxant, which Joan told me I used to take for backaches and always made me feel sleepy. Although this medication made me feel more rested, it didn't work either. Even after taking two or three of the little yellow pills, I was still wide awake, leaving me tired *and* lethargic, which was no improvement over just plain tired.

When I used to run the jet charter business, Joan said, I often got calls in the middle of the night from brokers or organ transplant teams that needed to fly doctors across the country. When this happened, I'd just start my workday because I couldn't get back to sleep, taking advantage of the quiet time at the office. But now that I didn't have a job to go to, I was left with nothing but my own ruminations over whether my memory would return or my insomnia would ever resolve.

Although I found it easier to deal with the headaches than this ridiculous sleep schedule, the pain had not let up either, and the two were interrelated. These days I usually had only a four-hour window when I was pain free, and the nearly constant pain was easier to handle in the evenings and on weekends, when I had my family to distract me. We'd go watch Taylor cheer at high school football and basketball games, and we'd visit her at Nando's restaurant, where I made her bring me special items such as chips with barbecue sauce. But when the pain woke me up—and kept me up—in the middle of the night, I felt miserable and alone.

During those long wee hours of the morning, I got sucked back down into the vortex of fear, anxiety, and uncertainty, which was even deeper now that the window had long passed for when the doctors predicted I would return to normal. I longed to sleep and for the pain to stop, wondering if this torture was going to last forever. Still haunted by the black hole where my knowledge used to be and the nagging lack of a diagnosis, I felt lost. I still didn't know what my values were or what I stood for; my sense of identity felt very soft, like a baby's skull, and my self-confidence was next to nil. When my anxiety led to panic, I climbed back into bed after Joan left for work, where I hid and cried for hours, feeling like a wounded bear.

Dr. Lanier finally prescribed the antidepressant Cymbalta for me, which helped to ease the panic and depression, but I still had those days where I couldn't drag myself out from under the covers even to watch television. It was difficult to find anything positive in my life, and sometimes reality was simply too much to face.

15

GRANT HAD BEEN DOING BETTER for the past few weeks since he got out of detox, or so he'd been telling us, although he still called with money-related and other life crises. Meanwhile, I'd been watching a lot more of *Intervention* and *Celebrity Rehab*.

But it turned out that Grant had been lying to us again. He called us in mid-March with the news that he'd been continuing to use since his last bout in detox and wanted to go back to rehab.

"I need help," he said.

Joan had been calling around, looking for a decent free or affordable rehab facility for Grant, but we couldn't find one nearby without a weeks-long waiting list. We were, however, able to get him into Ocean Hills Recovery, a ninety-day residential program a former counselor had recommended, which we hoped would give him a better shot at stability. The only thing was that it was in Dana Point, California, about thirty minutes north of Oceanside.

Putting some distance between Grant and his dealer and drug-using buddies in Arizona was Grant's choice, but to me it sounded like he wanted to run from his problems and avoid dealing with the reasons he took the wrong path in the first place. Either way, Joan and I were tired of all the hassles and hoped that after he finished treatment he'd find a job and an apartment there with his new, sober friends.

This would be his third stint in rehab, and his habit was getting pretty expensive. He'd spent a couple weeks in a failed outpatient program for cocaine addiction in October 2007, then two weeks later he did a six-week stint in a residential program. Every time he relapsed, it angered me even more that he'd let us down—again.

I can't understand why he can't stop the lying and doing drugs. Can't he see how it's hurting his family?

But if I was tired of his lies after just a few months, I could only imagine how Joan felt after dealing with them for nearly two years. For me, the hardest and most confusing part of this was my inability to counter my anger toward him and my self-doubt about my parenting skills with memories of the good times Joan said that he and I had shared while he was growing up. All I had to go on was the irritation of the moment.

. . .

Grant stayed the night at our house under close watch before he and I headed to California at 6:00 A.M. on March 18, as Joan was leaving for work.

"You have to get better," Joan told him tearfully. "This is yet another opportunity. We're spending a lot of money on this. You have to give it your all. You can't live like this—you will die."

As we headed toward Interstate 8, the freeway that led us into southern California, Grant fell asleep. This was okay with me because I didn't have anything more to say to him. At least by taking him to rehab, I could know that he was safe and getting better, giving Joan and me a much-needed respite from worrying whether he was getting high, getting hurt, or, even worse, hurting someone else.

Grant slept for most of the trip, waking up at the halfway point in Yuma for a bathroom break and a drink.

"Is there anything else you haven't told me?" I asked, hoping to avoid any more nasty surprises.

"No, no," Grant insisted. "Nothing."

Once we arrived at the Oceanside harbor, I put him to work washing down the boat while I worked on the interior, monitoring him carefully to ensure that he didn't take any shortcuts. Joan and I had decided that I would keep Grant on the boat with me one night then take him to rehab early the next morning.

We went to a Mexican restaurant to grab some tacos and burritos, where Grant told me about all the times we'd come out to California as a family before my accident, traveling to many motocross events even before we'd bought the Meridian.

"We had so much fun with the Jet Skis and just hanging out on the boat," he said, describing the leisurely excursions to Catalina Island and Huntington Beach on the Fourth of July and my birthday.

Then he grew somber, saying he was scared about the months and years ahead. He seemed like a different kid, acknowledging that his days of participating in family vacations would be over for quite some time. It saddened me that he saw his future in such a bleak light.

I tried telling him that his life wasn't finished, it was just beginning, and that he needed to start a new life without drugs. I wanted so much to hold and console him, but I had learned from my TV education on addiction that because nothing else we'd tried had worked, what he needed most from us now was tough love. If Joan couldn't give it to him because her emotional attachment was too strong, then I was going to do it on my own. I was the logical choice, and even though I didn't really know how, I certainly would give it my best.

"Your days of fun are over for now," I said. "You need to straighten out your life before you can think of having fun because fun is what got you into this mess to begin with."

We spent the rest of the night watching hockey on television. After being out in the sun all day, we were both worn out, so around 11:00 I told him to go to bed. My plan was to leave for breakfast at 7:00 in the morning.

In the meantime, however, I was worried that he might try to sneak off in the night to get high. "If I hear you moving around, I'll break your f---ing legs," I said.

Thankfully, he was still there the next morning. We gathered up his belongings, making sure he didn't forget anything, had a quick bite, then headed north on Interstate 5.

The Ocean Hills treatment facility was a two-story blue and tan house in a middle-class neighborhood´a few blocks east of the highway. The place seemed a little weathered, with a yellowing lawn, but it matched the rest of the neighborhood. The owner, Shaggy, a man in his midforties with bleached brown hair that hung just past his shoulders, looked like a beach bum. He and Thurman, the intake counselor, greeted us when we arrived for our 8:30 appointment.

Thurman was a muscular, hard-nosed guy with a shaved head who reminded me of a marine. He gave us a quick tour of the place, which had room for twelve residents, ages eighteen to twenty-five. Including Grant, it housed ten, four of whom were women, which, as you might imagine, could make for some obvious challenges. But it wasn't so bad—the back wooden deck had an ocean view.

The office was in the garage, where the door was always open. The residents often smoked in the driveway and on the back patio. We sat out there, filling out the necessary paperwork, while the counselor asked Grant some questions.

"When was the last time you used drugs?" he asked.

"Five days ago," Grant said.

I noticed that Grant's speech was slow and lethargic, and so was his thinking; he was struggling with questions that even I could answer.

"What the hell's the matter with you?" I asked him.

"I'm tired," he said. "I didn't sleep well."

"I don't care if you're tired. Pay attention and talk right."

I was determined to be hard-nosed with him, giving him my best tough-love act. After I'd put the $7,500 for the first month on my credit card, Thurman asked if I wanted to stick around while my son got settled in. But feeling a jumble of emotions, I felt the urge to get out of there as soon as possible.

"No, I'm going to take off," I said.

Grant walked with me to the end of the driveway, started to cry, and gave me a big hug. "I'm sorry for putting you guys through this," he said.

"Don't apologize," I said, hoping he could see how pissed I still was. "This is your chance to get better and to make everything right. Do your job here, and make sure you get everything out of it that they teach you."

I hugged him and kissed him on the check. "Good-bye. I love you and stay strong," I said.

"I love you too, Dad."

As I walked to the car, I fought back the tears, making it around the corner and down the next block before I could pull over and let the tears come gushing out. I felt horrible, dropping Grant at a depressing place like that, where we didn't know any of the people. I felt like a terrible father who had abandoned his son and left him to fend for himself.

I called Joan at work and shared my sadness with her. "I tried so hard to make it through saying good-bye without breaking down," I said.

She tried to calm me down but mostly just let me vent. I told her how disgusted I was with myself, being such a hard-ass with him on the drive to the boat from Gilbert. By the time we'd gotten to the rehab place that morning, I said, "Grant looked like a beaten man. It was so hard because my heart went out to him, but I was still just so angry with him."

"Look, I know this is tough," she said, "but we're doing the right thing for him and he really needs this right now. He needs us to be tough on him."

Joan always knew what to say to make me feel better, but I still felt like crap. Deep down I knew I could handle losing my memory and everything that came with it, but I wasn't sure if Grant could handle getting clean. That's what really scared me.

Can my son get through life without using drugs again?

. . .

After talking to Joan for an hour, I drove back to the boat, where I sat on the top deck for the rest of the day thinking and taking in the view of the boats, the people eating and drinking on the outdoor decks of the restaurants nearby, and the folks milling around on the boat ramps.

I pondered what the future held for me and my family and what I was going to do if my memory didn't return. As I tried to relax and get my mind off the headache that was pounding like a drum, I mentally prepared to spend the next couple of days alone until Taylor and Anthony arrived for spring break on Saturday. They were planning to stay until Wednesday, when we would caravan back together.

. . .

Later that evening I was watching a hockey game at Rockin' Baja Lobster, a sports bar, when Shaggy called to tell me that Grant was going through withdrawal, experiencing painful stomach cramps, alternating bouts of hot and cold, and an inability to sit still. "He needs to go to a medical detox facility," he said.

Apart from being furious, I couldn't even comprehend this. "How is he going through withdrawal when he hasn't used in five days now?"

Shaggy said that Grant had lied to us—he'd sneaked out of our house in Gilbert the night before we left for California and bought some heroin.

I was numb. I simply could not believe what I was hearing.

"He's going to be detoxing, and we're not set up for that," Shaggy said, explaining that there were medical risks, and they wanted to take him to a place that charged $550 a day, where he needed to stay for at least three days.

"I'll have to talk to Joan," I said. "I'll call you back shortly."

I was so disgusted with my son. If he'd just been honest, we could have taken him back to the free place in downtown Phoenix before having to throw even more money at the problem in California. When I told Joan that I'd asked him repeatedly on the drive out if there was anything else he hadn't told me, she was as pissed as I was.

"That f---ing liar!" she said. "Not only do we have to spend all of this money on rehab, now we have to spend $1,500 more on detox."

The more we talked, the higher the anger bubbled up inside me. "Why does he lie so much?" I asked.

"He's an addict," she said. "He's not capable of telling us the truth."

After discussing our options, we decided we had no choice but to send him to the detox facility Shaggy had recommended. I called him back, asking him to tell Grant that we didn't want to hear from him while he was in detox; we would call Shaggy to check on him. "I want him to go through as much pain as is tolerable," I said. "I don't want this to be easy for him."

"Okay," Shaggy said, as if he completely understood.

Because I couldn't do anything more at this point, I stayed at the sports bar, although I really wanted to drive up to Dana Point and beat Grant's ass. His drug addiction was costing us a small fortune, and yet he seemed completely oblivious to that.

I was so upset, my anger kept me up for two nights straight, and the pain pills did nothing to numb my headache. All I could do was watch television, cry, miss my wife, and try to process what was happening.

The nights were chilly, so I drank coffee to stay warm and dozed off during the day for fifteen-minute catnaps. I read more about addiction on the Internet and rehashed Grant's lies in my mind, which only served to prolong my fury. I needed Joan to help calm me down, and I wanted to help her do the same, but I'd promised Taylor that we could hang out on the boat during her vacation.

While I waited for her and Anthony, I passed the time talking to our boat neighbors, Barb, Ray, and Davey. I updated them about everything but Grant because I was embarrassed and didn't want to look like I'd been a bad parent.

When the kids showed up, they were more interested in Jet Skiing or Boogieboarding than spending time with me. I spent most of my days wiping down the dust and excess oil from the boat engine, preparing for the sale we hoped would be forthcoming.

After having dinner with them, I talked to Joan about Grant and us for hours at a time. Sometimes we realized we were both watching the same show, such as *Two and a Half Men,* and we shared a few laughs and talked during commercials.

I felt lost without my wife, almost as if I couldn't function properly without her. I'd heard the expression "he's my right-hand man," and this was truly appropriate to describe what she meant to me. But she wasn't just my right hand; she was my everything. When we were apart, nothing made sense. Given the circumstances, spending this time away from her was even more difficult because I wasn't able to hold her or feel her next to me. I missed her comfort.

Although I enjoyed spending time with Taylor and Anthony, Wednesday could not come soon enough for me. I was eager to get home to Joan and familiar territory because the boat only reminded me of Grant's drug habit and my anger. My sanctuary had become more like a prison, and the joy of being there was gone.

16

A FTER MY SUCCESS with selling our small fleet of cars, I moved on to my watch collection. Unlike the old Scott, who Joan said changed his watch almost every day, I'd become a creature of habit and had been perfectly happy wearing the same one for the past several months, which was normal for someone with a brain injury. My lightweight Citizen Eco-Drive Skyhawk had a comfortable black rubber strap, and, being solar-powered, it was also convenient because I never had to change the battery.

Reading online about the features and widely variable market values of my thirteen timepieces, I learned that the Skyhawk was a favorite among pilots because it told the time in forty-three cities worldwide and also calculated fuel time and flying speed.

Joan couldn't remember when or where I'd gotten the watches other than she'd given me the Chase Durer Trackmaster for my birthday. She did say, however, that every time I'd bought or sold an aircraft for a client, we'd go shopping—Joan for clothes or shoes and me for a new watch. But because none of them held special meaning for me now, I saw no reason not to liquidate them to generate some household income. Joan was doing her part; I wanted to do mine.

Building on my car sales experience, I researched the watches' wholesale, retail, and private sale prices, averaged them, then listed them on a legal pad along with what I originally paid. After deciding to keep three of them for a little variety—the Skyhawk; the Trackmaster, which had a stainless-steel strap, lit up at night, and contained a stopwatch; and the IWC, a dress watch with a leather strap—I crossed them off the list and asked Joan what she thought of it.

"Oh, you made up a spreadsheet," she said.

"What is that?" I asked.

"This form you made, comparing the cost of each watch, is called a spreadsheet."

But she still hadn't answered my question. "Okay," I said, "but what do you think of the list?"

"This looks good, but what are you going to do with it?"

Presenting my action plan, I said I would put ads on Craigslist and sell the remaining pieces to a retail store that sold used jewelry, aiming to get as close as possible to the current selling price.

Joan cautioned me to take safety measures so I didn't get robbed or scammed. Apparently we'd almost got caught up in a Craigslist caper in 2007 when we'd posted an ad to sell one of Grant's motorcycles. The buyer insisted on paying us with a $5,000 cashier's check for a $3,000 motorcycle, asking for the $2,000 balance in cash. Luckily, the bank determined the check was fraudulent before we completed the transaction.

Joan and I agreed I should meet potential Craigslist buyers at my office building to keep our home address secret.

I was amazed how many watches were for sale on Craigslist, and although I wondered if I had too much competition, I forged ahead. I listed three watches, including a new Rolex Explorer II and a barely worn Omega Seamaster Planet Ocean, both of which I priced at $4,000. My research showed that Rolexes held their value better than any of my other watches, so I was confident I could get my asking prices. The Omega, which I'd purchased for $5,550, had no scratches but felt like I had a boat anchor around my wrist.

I waited for the buyers to show up, but no one called for several days, and, even after the calls began to trickle in, I quickly grew frustrated with people lowballing my asking prices and failing to show up for appointments. I wondered if I was wasting my time. A week later I got a call from an ASU college student who was interested in the Rolex.

"The pictures look really good," he said. "I'd like to see it in person."

We arranged to meet within the hour in my office building lobby, where I knew they had security cameras in case he tried to rob me or accuse me of robbing him. When he arrived, he was in his early twenties, blond, tall, and well groomed in a polo shirt and khakis. In other words, he looked as if he could afford my watch.

As we sat on the black leather couches, he put on the watch and stared at it longingly as if he were trying to find a reason not to buy it. I wondered if that's how I'd felt when I'd bought it or if I'd been too spoiled to appreciate its handsomeness.

Finally he sprang to his feet, saying, "I want it, but will you take $3,500 for it?"

"No," I said, "I appreciate it, but I'm confident that this watch will sell due to the condition it's in." I started putting it back in the box, which seemed to prompt the young man to go for it.

"Okay," he said. "I'll pay the $4,000, but it's more than I wanted to spend."

After I told him I only took cash, he left to go to the bank, promising to return in half an hour. Meanwhile, I went up to my office to make copies of the sales receipt and registration for my records, and sure enough, he came back as promised. After he counted forty one-hundred-dollar bills into my hand, we shook on the deal, and he left with a big smile.

It was nice to see that this watch had brought him happiness when it wasn't doing anything for me but serving as a reminder of my previous excess.

Thank God I didn't owe any money on these watches. What kind of greedy man was I, needing to surround myself with all these cars and watches when I've seen so many people on the news going hungry, living with no phone or running water, not just in other parts of the world, but here in the United States too?

I was proud that I'd been successful enough to buy these luxuries, but when I considered how much I'd indulged myself, I felt nauseated.

How can there be so much difference between what some people have and others don't have? And how many poor people could I have helped rather than spending money on these items that I didn't need?

I wondered how I'd gotten so off track, trying to build wealth instead of focusing on what was truly important—my family. Was that partly why I'd missed noticing that my son had started down the wrong path in life? Maybe I'd been too preoccupied to see what was right in front of me.

Several weeks later I sold the Omega Seamaster to a man from Tennessee. After we talked on the phone, he agreed to wire the money into my account and trusted that I would send him the watch. He paid my asking price and an additional fifty dollars for shipping. When he received the watch, he called to tell me that it was in better shape than the photos had indicated, which seemed to be an unexpectedly pleasant surprise. "There's only one small scratch on the clasp," he said, astounded.

"I know, I didn't wear the watch," I replied.

"It's just very refreshing that somebody put a good product on Craigslist," he said.

As happy as I was to hear this, it made me realize that I needed to be more careful in the future. Before the accident, Joan said, I'd become bitter and untrusting due to some bad experiences in the business world. Right after the accident, I had felt frightened of people unless I knew everything about them, but now I'd become almost too trusting.

I was unable to sell the third watch on Craigslist, so I decided to take it and the other eight watches to two retail stores in Scottsdale and take the highest offer. By my calculations, the timepieces were worth around $22,000. I knew I wouldn't get that much; I was willing to take less if I could complete the sale quickly for close to that price.

A buyer at the Estate Watch & Jewelry Company offered me $19,500 for the bunch. I told him I needed to think about it and would call him later that day, then got in my car and drove to the other store, Scottsdale Fine Jewelers, which was about five miles away.

After chatting with the owner and his wife about the sad shape of the aviation business and my desire to convert the watches into cash, he offered me $20,800.

I stepped outside to consult with Joan by phone. "That seems pretty close to what we talked about," she said. "See if you can get a few hundred dollars more."

I went back inside and asked for $21,100, and we finally settled on $21,000.

"Okay," I said, "we've got a deal."

I had to say I felt relieved to get rid of these unnecessary items. I also felt a keen sense of satisfaction about pulling my own weight around the house and improving our financial security.

. . .

With our twenty-fifth wedding anniversary approaching, I decided it would be nice to renew our vows like I'd seen people do on TV. Joan had confided that she was worried I would wake up one day and want to run off with someone else because I couldn't remember all our years together. Knowing that my love for her had been growing stronger every day, I thought this gesture would help her see how committed I was to spending the rest of my life with her.

"Our twenty-fifth anniversary is coming up, and we're going to Hawaii. Do you still want to renew our vows like we planned before the accident?" I asked, expecting her to jump at the chance to do this during our upcoming vacation.

Joan got quiet and paused before she answered. She looked as if she felt backed into a corner and had to decide whether to hurt my feelings or be bluntly honest with me. "I don't feel that you know what love is yet," she said. "It might be a good idea to wait until you know what it means."

"But I do love you. I don't know how I felt before the accident, but I know how I feel now, and I know how I feel when I'm not with you, and this has to be love," I said, still reeling from the shock of what she'd said. "Why don't you want to remarry me?"

Reading my feelings of rejection, she grabbed me and tried to soften the blow. "It's not that I don't want to remarry you. I love you, and I have known you for most of my life, but you've only known me for a few months. I think it would be better if we waited until you know for sure that you still want to be married to me."

I told her that I may not know everything about her, but I must have loved her for the past twenty-eight years or we wouldn't still be married. I slowly realized, however, that she might be right. Maybe I did need to get to know her better, and we could renew our vows at a more appropriate time. She seemed to have guided me in the right direction up to this point. Who was I to start doubting her now?

We did end up going to Hawaii, but we didn't renew our vows. Instead, she spent our anniversary telling me, minute by minute, what we'd been doing on our wedding day twenty-five years earlier, until we fell asleep in each other's arms that night. I could feel my love for her growing stronger every day.

. . .

Now that I'd sold off what I could, I wondered what else I could do to help out financially. We were still hoping that my memory would return, but even if it did, I wasn't sure I would be able to pick up the jet business again. After all my work to make it successful, the company still had value, even without me at its helm, so in the worst case, we figured we could sell off its assets and client base.

In March we managed to get out of the last five months of our office lease, which saved us $5,000 a month, but we thought it best to keep the business alive in some form. So we switched over to a virtual office service for only $200 a month, which gave us a business address, phone and fax services, and a place to meet with clients if necessary. Calls to our business number were still answered by the same reception desk, but they were now relayed to voicemail; faxes went to my eFax, which I could access from home or anywhere else I could get on a computer. At first Joan handled all the callbacks because I didn't know enough to talk to anyone. When I felt better, we decided, I would start calling some people myself.

I was desperately looking for a way to take control of my recovery and my destiny, find a way to empower myself, and develop a plan of action—to find a feeling of purpose when I looked in the mirror each morning instead of the hopelessness, despair, loneliness, and helplessness that stared back at me. Knowing how bad and alone I felt and how much I wanted help but couldn't find it, I figured there must be others out there who felt the same way— people who had suffered brain injuries or other traumas.

Maybe my pseudocelebrity status of having played in the NFL and becoming a successful aviation entrepreneur will give me an entrée to get people to let me help them.

Although Joan kept telling me how successful I'd been, I was still struggling to comprehend what "success" meant and what it felt like, along with many other emotions I didn't understand and had to ask Joan to put a label on so she could help me figure out what I was feeling and why.

"What is success?" I asked her.

"It has many different facets," Joan said. "Some people relate it to money. Some people relate it to having a healthy family, happiness, raising good children, being productive in society, and being with the one person you love."

Knowing that we'd done charity work to help others in the past, I hoped that this newly revealed path might create a sense of success in me, which could then lead to further success.

Why not help others and myself at the same time?

"Maybe this happened for a reason," Joan told me. "Maybe God has a purpose for you, and this is it."

I still didn't understand what God was, but for now, that sounded like a pretty good affirmation. As I searched for a new career, I vowed to be more conscious of how I could help others in whatever job I chose. I only wished I knew how to help my son.

17

U NTIL THE INVITATION showed up in my email inbox, I
didn't realize I'd been a member of the NFL Alumni Association's Arizona chapter since 1995. The group was at-
tempting to revive itself after being dormant for four years and was
planning to meet.

"Do you think I should go?" I asked Joan, showing her the
email.

"If you want to check it out, go ahead," she said, trying to be encouraging without pressuring me into going somewhere I wouldn't
feel at ease. She'd figured out that I was more comfortable in a
room full of strangers than in one filled with people I was supposed
to know, and because I wasn't acquainted with these particular
players before the accident, I'd probably be okay.

It sounded like a good opportunity to get out of the house for a
few hours, but more important, I'd have a chance to connect with
men who had shared similar experiences playing football in college
and the NFL—smelling the same dirt, suffering the same injuries,
and experiencing the same successes and failures. How could I not
want to meet these guys, and what better way to learn about playing professionally than by listening to their stories? I was curious
to see how they acted and dressed, what they did for a living now,
what they ate, and what cars they drove. Any information that
helped me progress in my new life and gave me a better perspective

on my old one was very important. It was also a possible avenue for gaining access to something I'd seen on the news—brain studies of former NFL players who had suffered multiple concussions.

That said, I must have changed my mind at least a dozen times, torn between wanting to go and fearing that if I did, a neon sign would flash across my forehead: "I'm stupid; I lost my memory." But the desire to explore this part of my past ultimately won out.

When the day in late June arrived, I set off for the Ocotillo Golf Club in Chandler, about ten miles from the house. Armed with my MapQuest directions, a notepad, Legendary Jets business cards, and a pen, I was mentally prepared to try another new experience. Having learned that less than 2 percent of all college football players get drafted, I only hoped that I would be accepted as one of the members of this elite club who had not only been drafted but also played on the field.

I arrived fifteen minutes early as usual, wearing a business casual outfit of golf shirt and pants, and found my way to the conference room. Standing outside, I took a deep breath and walked in cautiously, hoping to sneak in and take a seat in the back. Instead, I was immediately greeted by Lisa, the association secretary, who gave me a big welcome and asked for my name and the team I'd played for. Giving me a name tag, she introduced me to Aaron Gersh, the association president, who had played for the Kansas City Chiefs. He and I chatted for a couple of minutes until he excused himself to greet some of the other men.

Starting to feel uncomfortable, I grabbed a Diet Coke and sat at the far end of the long conference table, away from everyone else. As a few other players came in, some dressed in shirts and ties from work, others in shorts and polo shirts, I saw them hugging each other. Lisa must have sensed that I felt out of place because she came over.

"I want to introduce you to another Cleveland Brown," she said, leading me over to an African American man a few inches shorter than me. I could feel the sweat beading on my face and my heart starting to race as she introduced me to Ray Ellis, who before his years with the Browns had played for the Philadelphia Eagles. Ray shook my hand firmly.

Preparing for this meeting, I'd spent a lot of time reading up on the NFL, the rules of football, the names of the big players from my time in the league, the names of the teams and their mascots, and the Hall of Famers. And yet we'd chatted for only a minute when he posed a question I hadn't anticipated.

"Which coach did you play for, Sam Rutigliano or Marty Schottenheimer?" he asked.

My mind went blank. I started to shake, and the sweat was now dripping profusely. I was confronted with the very dilemma I'd feared most. Any moment he'd read that sign on my forehead. My fight-or-flight response fully engaged, I came up with an escape.

"Hold on to that question," I said. "I'll be right back to answer it." I fled into the men's room across the hall, where I hid in one of the stalls, shaking.

What the hell am I doing here? I'm not ready to be in the real world yet!

As I tried to pull myself together, I looked in the mirror and told myself that I had made it there, and I needed to stay. The embarrassing moment was over. I'd already looked stupid. What more could happen?

I washed my face, ran a wet towel across the back of my neck, and wiped the sweat from my arms. Having recovered my composure, I walked back into the meeting, which was just starting, and took a seat. By this point ten of us were sitting around the table, with room for maybe ten more.

Aaron explained that the alumni chapter had been largely inactive due to lack of interest, and its fund-raising efforts had been combined with those of the Arizona Cardinals. Aaron, who wanted to make the group independent again, said we should be able to rebuild to be even stronger than before because three hundred fifty retired players had homes in Arizona. The group's mission was to raise money for charities geared toward helping children, and this year it was working with the Children's Miracle Network, which raised money for children's hospitals.

Now that really piqued my interest. Although the group also addressed the issue of pensions and health benefits for retired players, its main goal was to organize fund-raising events such as golf and poker tournaments for these charitable organizations. To me, there was nothing more precious than children. The ones from broken or poor families, who didn't have enough money to buy medication or fend for themselves, well, they just tugged at my heart.

Joan told me that when I'd been part of this group back in 1997, I'd helped with one of its tournaments but had subsequently lost interest. I wrote checks to help out, but I didn't know how else to contribute because I was too busy. Now that I couldn't do my previous job and was waiting for the headaches to subside while I learned the world and searched for a new career, I had plenty of time, and I wanted to use it to help others.

One of my other fears about coming to the meeting was that I would walk into a room of superstars, and after playing professionally for only a few years, I would be a nobody in comparison. But as it turned out, Aaron had had a similar experience. He played a year as a linebacker until he blew out his right knee, went on the disabled list, and had seven surgeries. In the end my fears proved unfounded; we were all there to help children.

As we went around the table, each of us introduced ourselves and said what we did now for a living: Aaron had become a hospital administrator after earning a Ph.D. in organizational psychology; David Recher, a former player for the Philadelphia Eagles and Minnesota Vikings, now sold ads for Clear Channel Communications; Floyd Fields, who had played for the San Diego Chargers, worked for Ruth Enterprise, a software company; Thron Riggs was long retired after playing with the Boston Yankees before the NFL even existed; Kwamie Lassiter, who'd played for the Chargers, St. Louis Rams, and Cardinals, had his own radio talk show, as did Ray Ellis; Larry Wilson, a Hall of Famer who had played for the St. Louis Cardinals and went on to be its general manager, was now retired; and Jenn Bare, a former Rams cheerleader, was a business manager for a global outsourcing company.

During the ninety-minute meeting, I didn't ask a single question but listened intently and took copious notes. Aaron asked us to consider taking on board member obligations to fill four positions: directors of youth, sponsorship, tournaments, and membership. Afterward, I stayed and chatted with the others, enjoying the camaraderie that they shared. It made me want to be part of this brotherhood even more.

Aaron and Kwamie welcomed me with their kind and generous words, helping me feel that I could make a home here, start doing something with my life, and rebuild my sense of self-worth.

When I got back to the house, I told Joan how stupid I must have looked not knowing the name of my coach. "Maybe I'm just not ready for this yet," I said.

But Joan shrugged it off. "So what? You didn't know an answer. These guys have been banged up enough; he probably didn't even remember he asked the question."

That helped, but I still felt like an idiot. I Googled the 1985–86 season for the Browns and learned the right answer was Marty Schottenheimer, a name I would never forget again.

After taking a few days to ponder the skills required in the open positions, I talked with Joan about becoming membership director. It seemed like a job with duties I could handle—following up with players who had let their memberships lapse and persuading them to rejoin. Unlike the youth director position, where I'd be expected to design programs such as football camps, or the sponsorship director, where I'd have to try to bring in big company sponsors and vet the charities that wanted our money, this one seemed to fit my particular and limited capabilities. Joan was thrilled that I would consider taking on such a big role. She said I was a great choice for this post, and it would also help me grow as a person.

The next day I called Aaron and told him I wanted to be considered for it. Still unaware of my accident, he said the experience of running my own aviation company would be very helpful. "I'd hoped you would step up to the plate and assume one of the positions available," he said. "I think you will do an excellent job."

At a meeting the following week, a few more players turned out, and we started planning our next golf tournament for January 2010. We took care of some housekeeping items, then it came time to vote on the board positions. After a unanimous vote, I became the new membership director.

I felt great, but it was quite typical for me to bounce between emotional highs and lows. One minute I would feel like I did now—on top of the world, as if I'd crossed a major milestone and taken a giant leap forward in my redevelopment—and the next minute I could feel that the world was on top of me. Knowing that, I was determined to push through and do the best job I could, just like the old Scott and one of his heroes, NASA flight controller Gene Kranz, who helped save the astronauts on Apollo 13 and whose memoir, *Failure Is Not an Option,* sat on my home office shelf.

This NFL alumni group was going to help me make a difference *and* learn about my past—how this game of football had created the person I was before the accident and, more important, how it was going to shape the new person I wanted to become.

Joan was proud of me when I came home with the news. "That's going to be awesome," she said. "It'll get you out, it'll get you connected."

We agreed that the NFL Alumni Association was going to help me regain the confidence I needed to start becoming productive in life and business again.

. . .

In spite of the successful steps I'd been taking to move on, my desire was undiminished to obtain a diagnosis and prognosis for my unresolved memory loss and my unrelenting headaches and insomnia. To say that my patience was wearing thin was putting it mildly.

Joan had done plenty of research into various tests and treatments and had requested a single photon emission computed tomography, or SPECT, scan several times to no avail from the three neurologists I'd seen. The test, they said, was too costly, controversial, inconclusive, or unnecessary because they thought my memory would come back.

When Joan asked the specialist we'd seen most recently for the scan and a functional MRI, he talked about referring us but ended up sending me back to a psychologist instead. Tired of people telling us that my memory loss was all in my head, as it were, we finally asked my primary care physician, Dr. Lanier, to refer us to a neurologist specifically for the purpose of administering this test.

The way it was explained to me, a SPECT scan was like a high-powered MRI that measured the amount of blood flowing to the different parts of the brain and showed any irregularities, including any reduced flow to injured areas, through a spectrum of bright colors. The technician injected a radioactive dye into your arm, waited thirty minutes for the dye to reach your head, then performed the test.

After refilling my pain medication, Lanier suggested we see Dr. Fern Arlen, a neurologist in Scottsdale for whom she had the highest regard and whom she would contact personally on our behalf to explain my condition in detail.

. . .

These days, Joan told me, it was difficult to determine what I understood or what preaccident knowledge I'd retained. She couldn't be sure if or when I fully understood a transaction, could appreciate the value of what I was selling, or could comprehend the implications of my actions. But, in general, she said, my mental capacity appeared normal to outsiders because I could follow logic and reason. Although I couldn't focus on any one task for very long, my short attention span allowed me to learn about many different things in quick sequence.

Overall, she said, I was far more forgetful about the little things than I used to be. As I was going to change a lightbulb, for example, I might set down the new bulb somewhere in the house and forget where I'd put it. I would take my shoes off, put them halfway underneath the ottoman in front of my chair, then not be able to find them twenty minutes later, shouting, "Who moved my damn shoes?" Or I would leave the house on an errand and forget something I needed to complete the task. Now these mishaps may sound common or insignificant, Joan said, but the old Scott never had such lapses.

By the same token, Joan told me that I was far more accepting of these small failures lately than I'd been before the accident, although at times I still blew up when Joan got us lost while I was driving or when she forgot something after I'd rushed her out of the house. One time we were heading out to the boat and stopped in the Mission Bay area of San Diego, where I got very aggravated at Joan for not giving me proper directions because she didn't know where we were.

I tried to explain to her how this worked for me. "I need to know where we're going because I feel like I can't afford to make mistakes, and it builds up great anxiety when we get lost because I'm already feeling so lost to begin with," I told her. "So it's in my best interest to prevent getting more lost by planning ahead."

"Well, we used to enjoy being spontaneous," she said. "We could just go out driving somewhere in San Diego—or some other city we weren't familiar with—and look for a fun place to eat. But I understand that the more you know, the less anxious you'll be, so we'll hold off on 'discovery' trips for a while."

One thing I'd retained were my organizational skills. Joan said I still knew what I needed to know to accomplish a task, probably after Googling it, and was quite capable of developing an action plan, as I had with the cars and watches. Where my memory could fail me was during certain steps in the execution.

There were still big gaps in my general knowledge. Joan said I often got a blank expression when we were talking and I had to ask her to repeat a word or phrase that I used to know. Sometimes I didn't understand the meanings of less common words and

needed her to explain them as well. Some everyday phrases simply didn't make sense to me, such as "can't see the forest for the
trees," "cut from the same cloth," "don't cry over spilled milk," and
"don't put all your eggs in one basket." If she said one of these and
kept talking, I missed everything that came after it until I got an
explanation.

These days I often went back and forth between the family
room and my office to look on the globe for a country I'd seen
on the news, such as Afghanistan, and determine where it was in
relationship to us.

When a complex subject such as a mortgage came up, I Googled
the definition to learn how it might apply to me or our family, then
went through our household financial documents to learn more.
I did the same thing when I heard terms on the news about "the
federal reserve," "a thirty-year adjustable rate," "refinancing," or
"loan modification."

Same-sex marriage was in the news a lot these days, particularly
with the approaching election, so I discussed this issue with Joan.
Before the accident, she said, we'd both felt that marriage should
be between a man and a woman, but if a gay couple wanted to
marry it was their business. I'd seen a lot about gays on TV, and
although I still didn't have a problem with them getting married,
it made me uncomfortable to see two men kissing. That was one
more thing that hadn't changed.

. . .

I made an appointment to see Dr. Arlen in late June, and because
Joan was working, I went alone. This wasn't a big deal for me anymore. I was used to seeing doctors by now.

Arlen was fairly tall with graying hair, a quiet manner, and a
stone-faced expression. She seemed to be the most knowledgeable
of the neurologists I'd seen to date, and she ran a series of question-
and-answer tests similar to those in my earlier neuropsych exam.
Upon reviewing the original results, she said she wanted me to

do a full exam again to rule out early dementia, even though she doubted this condition was in play.

After we'd talked for more than an hour, she seemed surprised when I told her that our past requests for a SPECT scan had been rebuffed. She agreed that we should order one and said she would ask a radiologist she respected to interpret the results.

She seemed puzzled that my memory still had not returned now that six months had passed since my fall. When I asked if she had any theories, she said, "I have no idea, but the SPECT scan should give us some answers."

I felt more encouraged after our appointment than I had in quite some time.

. . .

On July 16 I went to another NFL alumni meeting and chatted with Lisa before the other folks arrived.

I was feeling more comfortable with these people, and if this organization was going to become a home for me, I wanted to be up front and honest about my condition. I also felt I could pick no better person to tell than Lisa, was who about my age and was very easy to talk to. She'd been so warm and kind to me since the start, and I was hoping that if she saw me struggling again she would step in and act as a buffer—in other words, be my Joan at the alumni group.

So I shared my tale with her and was shocked when she started to cry. "You are one of the most courageous men I've ever met, to be able to survive and adapt the way you have," she said.

"Looks are deceiving," I said, explaining that I rarely left the house and took a big chance by coming to the initial meeting.

Lisa informed me that her husband, Bob, was the chapter's historian. "I'm sure he would like to hear more about your story," she said. "He might even write a feature on it and submit it to the NFL headquarters to try to get it published in their monthly magazine."

I told Lisa I would be open to that. It felt good to unburden myself of this secret I'd been carrying around for months, but I also asked her to keep it to herself and Bob for now.

"Absolutely," she said, "and thank you for sharing that story with me. It is truly inspirational."

18

J OAN WAS SITTING IN MY LAP one night as we watched one of our favorite movies, *The Bourne Ultimatum,* when the phone rang. It was Grant, calling collect—and drunk—from a pay phone, just a few days shy of reaching ninety days of sobriety. Joan put him on speakerphone so we could both hear him tell us that he'd been kicked out of rehab.

"I'm glad that you're okay, but what the hell is wrong with you?" I asked.

He didn't have a good answer for me. "So what are you going to do?" I asked curtly, feeling the rage rising inside me.

"I don't know."

"Well, you'd better figure it out," I snapped, "because I'm not going to figure it out for you."

"I'll call you later. I'll be fine."

Joan started to cry, but I was so furious that I was hoping never to hear from my son again. Somehow I managed to put my feelings aside and help Joan deal with hers because she was getting hysterical. I took her in my arms and tried to console her. "We'll figure this out, I promise," I said. "He's going to be okay."

I really didn't know if this was true; I was just trying to calm her down so we could come up with a plan. This roller-coaster drama with Grant was making both of us crazy.

"He's never been homeless," she said. "What if he overdoses because no one's there to help him?"

"You have to relax," I said. "He's going to be okay, and I'm sure he'll call us soon."

As we lay awake, our emotions went from dashed hope to worry, panic, and, at least for me, anger. "I hope he gets arrested so he has some consequences for his behavior," I said.

Joan wouldn't go that far, but she agreed he should feel some ramifications of his actions. For the moment we were the only ones feeling the pain after shelling out all that money for failed rehab programs.

I tried to reassure her that Grant was a strong tall kid with tattoos—big enough that no one would mess with him—so I wasn't worried about him being homeless in a park or on the beach for a couple of days, even a month.

"He's a resourceful kid," I said. "But he's also not that mentally strong to be able to live like that for long. He'll figure out a way to make other arrangements."

I dozed in and out of sleep all night, with Joan nudging me every so often and asking, "Do you think he's okay?"

I kept trying to console her, but she wouldn't let go of the what-ifs. "What if something happens to him?" she asked, crying. "What if he gets so lonely that this time he takes his own life?"

"I don't think he'll do that because, deep down, I don't think he wants to die."

"You don't know that for sure," Joan said.

As night turned into morning, we felt helpless, still hoping to get a call from him and not the one we dreaded most: "I'm sorry to inform you that your son is dead."

This was no way for us to live. I was strong enough to deal with my amnesia, but this on top of everything else was almost too much to bear. Sometimes I wished Grant was out of our lives so I could concentrate on getting better and taking care of Joan and Taylor, who actually wanted to be part of our family.

. . .

Grant finally called collect around 10:00 that morning, saying that he and one of the girls who got kicked out of rehab with him had drunk some vodka and slept in a park. He said he would call us again to let us know that he was okay. We both told him we loved him and that we wanted him to get help.

"I love you guys too," he said, and hung up.

Joan was beside herself. She started calling counselors from Grant's previous programs, his new sponsor, and his old sponsor in Arizona, asking for advice. They all said the same thing: "We can't help him anymore. He has to want to get help for himself, and if he wants it he knows where to get it."

This seemed to calm Joan enough that she could move back into tough-love mode.

. . .

We didn't hear from Grant until later that evening, when he said he was getting high again and going to stay at his friend Justin's apartment in Oceanside. Justin (another friend of Grant's, who we called "Justin from Cali") was the son of one of Joan's friends, whom Grant knew from motocross. Joan did not think that Justin was a drug user, and we were relieved to know that he wasn't sleeping on the street.

Over the next few days Justin kept calling Joan, worried that Grant was going to overdose in his apartment on the heroin that Grant had bought with money one of the girls had panhandled at a gas station.

"Kick him out," she said. "You can't help him, and I don't want you in the middle of this. He needs to get help on his own."

Joan and I became more worried when Grant called back a couple days later with an attitude that frightened us both. "I don't care anymore. Maybe that's all I am, a drug addict, and always will be. I just don't care," he said. "I have enough money to go buy enough heroin to kill myself."

Joan began to cry again. "Please don't do this to us. We love you so much. You need to get back into a program. You need to call your sponsor."

"Why call him? He's just going to try to talk me out of it. All I want to do is get high, so who cares?"

"We do," Joan said pleadingly.

Grant left the threat hanging in the air and abruptly ended the conversation. As a father, I wanted to drive out to California to make sure he was okay and to get him some help, but as a man I wanted to grab him through the phone and smack him.

Through Joan's investigative efforts, she found a sober living home called Donna's House in Orange County. The place had a good reputation and so did Amy, the woman who ran it, who, we were told, wouldn't put up with any nonsense from Grant. We took comfort in knowing that if he wanted to get help we'd found him a place to go for $800 a month. It seemed like a better option than bringing him back to Arizona.

Grant eventually called back, saying he had no more money. The girl had gone back into treatment, and he was hungry because he hadn't eaten in two days. *Now* he wanted our help.

I had Domino's deliver a couple of pizzas to Justin's house, and Joan told him about the sober living program. Grant said he wanted to go the next day.

"Then it's up to you to call Amy and arrange it," I said. "We're not going to do this for you."

"I'm really sorry for all I've put you through, and I really don't want to live like this anymore," he said. "I love you guys."

The next morning, as Grant started yet another stint at getting sober, we found ourselves feeling hopeful once again that it would click for him this time.

. . .

In the meantime, I'd been popping Percocets like they were Skittles because it now took nearly twice as many to relieve the pain. That meant I sometimes had to take twelve a day. With the increase, Joan said, she'd noticed that I seemed generally more agitated, with more sudden angry outbursts.

"Do you really think it's the medicine or the fact that Grant is causing both of us more stress?" I asked.

"That's a possibility," she replied, "but I know what you're like on pain medicine from previous surgeries."

"Okay, I got it."

I realized that Joan was giving me a not-so-subtle hint that she wanted me to start controlling my anger, which I hadn't even realized was a problem. We decided I shouldn't be taking so much Percocet, and I certainly didn't want to become dependent on prescription medication, so I made an appointment with Dr. Lanier to ask about other options.

Lanier recommended that I switch to a low dose of an extended-release form of oxycodone called OxyContin and take it three times a day to keep a constant dose in my body. This, she said, should avoid the ups and downs I was getting from the Percocet, but I would still need the oxycodone for the breakthrough pain.

When I explained to Lanier what was going on with Grant, she said she believed that additional stress rather than the medication was more the likely cause of my angry outbursts. She acknowledged how stressful my amnesia had been on both Joan and me and said she couldn't imagine how we were doing as well as we were.

"Looks are deceiving," I told her. "I'm finding it hard to make it through some days without just lying in bed and crying all day."

Lanier, who had been wonderful to both of us, was really the only physician who seemed to take a personal interest in helping me. It seemed that she and her nurses had done everything they could to make this condition bearable.

. . .

The weeks dragged on as we waited for the results of my SPECT scan while Dr. Korn, whom Dr. Arlen insisted was the only one who could read the results, was away on vacation. With each week that passed, I felt more drained. The tension was taking its toll.

It was an entire month before I received a call from Arlen's office on July 19. I made an appointment for the next morning, and for the rest of that day conflicting thoughts raced through my head.

First I felt positive, excited, and hopeful: *Finally, this could be it. I will find out why I'm still without my memory. Maybe, just maybe, they'll find it's something so simple that I can take a pill for it or undergo some sort of treatment to make me feel better.*

But then the fear and negativity set in: *This is just going to be like the MRIs, the CT scan, the EEG, and every other test I've had so far that showed nothing. Maybe the doctors are right. Maybe this is all in my head and I am crazy.*

When Joan got home that evening, I told her I was scared but I hoped this test would show something definitive. "I really need to know why I can't remember my life," I said. "Not knowing just might make me go crazy."

Joan hugged me and said, "No matter what this test reveals, we are going to make it through this. You're very strong, and you have the love and support of your family. We're always here for each other."

I wondered if in my former life I'd had the same ability, to always know what to say to make her feel better.

I didn't know why, but I felt like I was going to wake up the next day and have a new life. Needless to say, I slept only an hour or two, wondering what the test results were going to reveal.

. . .

Joan left for work early the next morning, saying she was eager to hear the news as well. "Good luck, and you'd better call me the minute you get out of that appointment," she said. "I love you."

"I love you too," I said, "and I promise you will be the first to know, good or bad."

I was already showered and ready to go by 8:00 A.M., and the minutes seemed to crawl by. I kept looking at the clock to see if it was time to leave for my 10:00 A.M. appointment. I arrived thirty minutes early, only to wait in the lobby until 10:15.

When the waiting room door opened, it was Arlen herself, and she wasn't smiling, so I immediately thought the worst. "Scott, come on back," she said.

We both sat down, and she told me she had the results, which Korn had sent over in a detailed report.

I couldn't wait another minute for them. "Okay, I have to know now, so please tell me what they are," I said, practically shouting with impatience.

Speaking in her usual unemotional monotone, Arlen said, "Based on the findings, it seems that you have a decrease in blood flow to the frontal and temporal lobes of your brain, and it appears that it was caused by a traumatic event, which in this case was your fall," she said.

"So does this explain why my memory has not come back?"

"The brain is very difficult to understand, but we do know that the long-term memories are stored in the temporal lobes, and the blood is definitely not flowing normally to that area, so yes, this is why your memory has not come back."

The scan, which showed healthy blood circulation in orange and decreased flow in blue, reflected a "dramatic reduction" in blood flowing to the front part of my brain, resulting in atrophy and a form of retrograde amnesia that was very uncommon, Korn's report said.

"Will it come back, and if so, when?" I asked.

"I don't know when it will or if it will," Arlen said. "All I can tell you is that the longer it takes, the less likely it is that you will make a full recovery. But if you do get memories back, they will start from your earliest memories and then work up to the later ones. You may only get up to age twenty or thirty; we just don't know."

"What happens if I don't get my memories back by, say, a year, two years?"

"If you don't get your memories back in a year and a half, chances are you'll never get them back."

MY LIFE, DELETED   193

That last statement floored me. "Wow," I said.

Asked if any treatment was available to increase the blood flow, she said, "No, unfortunately, there is no known remedy for this condition you have. Your memory will either come back by itself or it won't."

Arlen recommended that I start concentrating on forming new memories and looking to the future, in other words living my new life without thinking about the possibility that I'd ever get my old one back. This diagnosis and her advice were exactly what I'd been looking for. Finally I had concrete evidence that something was wrong with my brain, that my problem was not psychosomatic.

Her calm parting words were the most powerful and true. "Scott," she said, "now you have closure. Go live your life."

I thanked her and left her office with a thousand thoughts running through my head. But the loudest one was *This is it. This is all I have to go on for the rest of my life.*

After walking robotically down the stairs from her office to the parking lot, I sat in my car in the hot sun. I wanted to cry but I couldn't. *Why cry?* I thought. *I've already cried enough.* Besides, my brief sadness dissipated almost immediately and slowly morphed into a feeling of lightness and relief. The burden of the unknown had been lifted, and I realized I had no reason to be sad because I was free. I wasn't crazy after all.

I felt a smile creeping across my face; I didn't know why, but I couldn't stop it and I didn't particularly want to. Then it hit me. I now had the information that was going to allow me to move on with my life. I didn't have to sit around and wait for answers anymore. I finally had the answer to why I'd lost forty-six years of my life, how my life had been deleted in an instant.

I must have sat in my car for twenty minutes before I realized that I'd promised to call Joan. She picked up so fast she must have had her cell phone sitting inches from her fingertips. "I have no blood flow going to the frontal and temporal lobes of my brain, and that's the reason I can't remember my life," I said.

"So when is it going to come back?" she asked, clearly not as happy as I was about this news.

I realized then that although I'd received closure from the diagnosis, this was not what Joan wanted to hear. She wanted the old Scott back.

"She said it could take up to a year and a half to return, but if it goes past that then it might be permanent," I said.

There was silence on the other end of the line. "So it may never return," she said, her voice cracking as if she were about to cry.

"Yes, but it could. And I can move forward with my new life now that I know I'm not crazy. There is actually something wrong and they found it," I said, hoping she would join in my feelings of satisfaction and resolution.

I thought I could hear Joan crying, only she didn't let on, trying to sound like she was okay with the news because I was. But I was more than okay. I felt great. The prognosis may not have been what Joan wanted to hear, but for me it was just what the doctor ordered—and that was *closure*.

19

T HE DAY AFTER MY BIG EPIPHANY, I woke up feeling so good that I set off first thing for the Brain Injury Association of Arizona in Phoenix. I'd finally learned how to use the navigation system in my car, and it guided me to their downtown headquarters without missing a step.

I'd come across the organization during a Google search for doctors, therapists, and other resources for brain injury patients and had called the office. *I can't be the only person suffering from this condition. If I can find others, it would do us all good to share information and learn from each other.*

The group's executive director, Mattie Cummins, told me she'd heard of retrograde amnesia but she'd never met anyone who suffered from the condition. I told her that I might stop by sometime to chat, and she welcomed the idea, so I didn't think she'd mind if I showed up on her doorstep without any notice.

Mattie was a tall, perky woman who smiled frequently. In her late thirties with dark brown shoulder-length hair, she happily led me back into her office, where we discussed my accident in more detail. Trained as a social worker, she was very open, insightful, and extremely knowledgeable about brain injuries, so much so it almost seemed as if she'd had one herself.

"How do you feel?" she asked, posing a question that, curiously, no doctor had ever asked.

"Lost," I said.

"With a brain injury, that's very common," she said. "But you are so far ahead, functioning-wise, of so many people with brain injuries. You look normal, you walk normally, you're able to carry on a conversation. . . ."

"But I don't know who I am," I said. "I'm walking around in a fog. Not having any memories keeps me from making any decisions based on prior experiences."

I asked if she could recommend anything to ease the adjustment to my new life, and she suggested that I get some counseling. I'd heard that before, but I had an idea of my own.

"I came here to see how you could help me, but more important, I came here to see how I can help the Brain Injury Association," I told her. "I want to become involved, and I want to start telling my story to others."

Mattie looked stunned, then she conveyed just how surprised she was. I'd received such devastating news only yesterday, she said, and yet here I was, offering her *my* help? "It is unbelievably incredible to me that you're here today," she said. "I don't know exactly how, but I'm going to figure out a way to make you a part of the Brain Injury Association."

After talking with Lisa some weeks earlier, I'd realized that I wanted to tell my story to a broader audience, I just didn't know how to go about it. When I found the Brain Injury Association, it became clear to me that this was the way, and after receiving my news from the doctor, I figured this was the time. Now, with Mattie's help, I was going to make it happen.

. . .

A month or so later Mattie called to tell me about a sixteen-year-old girl named Taylor Ward from the East Valley of Phoenix, who had fallen in April, hit her head on the bleachers while playing volleyball for her school team, and ended up with the same condition as me.

I couldn't believe what I was hearing. Joan and I had been searching for months on the Internet for other people like me. I was amazed to learn that this young girl lived only ten miles away, and she too was probably wondering who she was and how she was going to make it in this world.

"Would you be interested in speaking with her, and kind of giving her guidance or reaching out to her parents?" Mattie asked.

"Absolutely. I would love to speak to her," I said. "That's what I want to do. That's how I think I can help."

Mattie said she would give the girl's parents my phone number and tell them to call me when they were ready. She told me to sit tight and wait for their call, explaining that parents of brain-injured children usually want to introduce help only if and when they feel it is appropriate.

I understood this, but I was eager to talk to this family and let this young lady know she wasn't alone, that I could be of some assistance by sharing how I'd coped with losing my identity and relearning the vast amount of information I needed to survive.

Mattie was brimming with ideas that day. She asked if I would also be interested in speaking with soldiers coming back from the war who had suffered posttraumatic shock and brain injuries from the IEDs, the explosive devices that terrorists had planted along roadsides in Iraq. She said I would be a good spokesman for the organization to help educate health care professionals and family members of the brain injured as well.

"People would be interested in speaking with you because you've played in the NFL," she said. "This would be good way to break down some of those barriers and help others overcome their disabilities."

"Yes," I said, adding that she could call on me anytime. "I'm more than willing to help in any way you see fit."

Mattie then asked if I'd be interested in becoming a member of the organization's board of directors.

"I would be honored to be considered and would do my very best to try to make a difference," I said.

She said I would be one of only two board members who were brain injury survivors, which made me feel very proud. With the connections I was making through the NFL alumni board and now this, I hoped I could somehow connect all these key people to make an even bigger impact.

. . .

Taylor Ward's mother, Kathy, followed up with me a couple of weeks later. "We would love to get together and for you to meet Taylor, but she isn't ready at this time," she said, adding that she hoped that day would come sometime in the near future. "She's mad because she has no friends. No one stuck by her. It's been a real ordeal."

I told her that I wanted to do anything I could to help her daughter heal, and she could call me when she was up to it or needed a friend. "You can contact me twenty-four hours a day," I said. "I'm up all night usually, so call me anytime, even if it's in the middle of the night. If anyone would understand what she's going through, it would be me."

Kathy seemed grateful for the offer. "My daughter can't sleep either," she said.

After speaking with her, I felt bad for the teenager with the same name as my daughter, and I looked forward to the opportunity to try to help her somehow.

. . .

I went to discuss Dr. Arlen's prognosis with Dr. Lanier, and she seemed somber when she entered the exam room. "How did you take the news?" she asked.

"I'm not quite sure why I'm okay with it, but I am," I said.

Once I told her I felt optimistic about the future, her mood seemed to lift. She then delivered a concurring second opinion, surprised and relieved that I was okay with it. Dr. Lanier said she and Arlen also agreed that I should see a therapist specializing in head trauma, who could help me improve my coping skills.

Asked how the Cymbalta and my pain meds were working, I said I felt much more even these days, had more control over my

emotions and headaches, and wasn't experiencing the same extreme lows as earlier in the year.

"I suggest you stay the course with these current medications
and try to stay strong," she said. As our family doctor, Lanier was
Joan's physician too, so she suggested that Joan seek counseling as
well so we would both be prepared to travel the long journey of
recovery and rebuilding still ahead of us.

. . .

By this time, I'd also sold our two Jet Skis. With the recession still
raging, we'd had some interest in the yacht but no offers, so we'd
lowered the price from $350,000 to $289,000 to $250,000.

The financial pressures were growing heavier by the day, and I
was feeling even guiltier that I was unable to provide for my family
the way I used to. I'd been doing my best to relearn my aviation
business, but after a couple of months I had to face the fact that
I wasn't going to master it with my limited knowledge base. That
said, I was determined to find a new way to make a living.

Over the past couple months Mattie had come up with an idea
that I thought just might work. "Why not start speaking publicly
about your accident and your amnesia?" Mattie suggested. I was
intrigued by this idea, which I'd never considered before.

One evening I broached the possibility with Joan. "What do you
think about me becoming a professional speaker and talking about
our story?" I asked.

Joan's first response was to laugh at the irony. "This is funny,"
she said. "You never would have gotten up in front of people to talk
before, let alone tell such a personal story about yourself and us or
a traumatic event that you'd gone through. But I think you would
be great at it, and people are going to be fascinated by this story."

I didn't really understand why I would've been reticent to tell
my story. It was so different from my thinking now. "Why would
I not want to speak in front of people? What's wrong with that?"

"You're just a very private guy, and you don't disclose much of
yourself or family to anyone," she said. "We had our company

picture taken, and you're not even in it." Nor would I go to net-working events, she said, choosing instead to send her or someone else to talk up the company to strangers.

"Well, I think it's a good idea," I said, "and Mattie said I can make a living at it."

Joan was starting to get excited about the idea too. "Yes, you can," she said, "and maybe we can speak together about how we've both gone through this."

The two of us wasted no time brainstorming ways to go about this. We both had stories to tell, and with Joan's previous experi-ence in setting up our companies' websites, she was well versed in how to proceed. She suggested I start blogging and Twittering, set up Facebook and LinkedIn accounts, and start journaling about my experiences.

I had no idea what most of those terms meant and felt somewhat overwhelmed. *Can I do this? What if I fail? Or worse, what if my story just bores people?*

But Joan helped me push ahead, educating me along the way, and together we set up accounts for me. Next, she hooked me up with a website developer named Kevin, a social media expert she'd met through a friend who had photographed our airplanes for advertising purposes. Although Joan was working full-time, she seemed excited to help me start my new career.

I met with Kevin several times at my office in Tempe. Once he heard my story, he was taken aback but inspired and took a per-sonal interest in getting me up and running. "People can relate to the inspirational story and all the trials you're going through," he said. "It's captivating, and people will want to read more."

. . .

I'd read blogs but didn't understand at first how writing one of my own would help me. After Joan and Kevin explained exactly how a blog could be intertwined with the social media I'd been learn-ing about, I was convinced as well. In the beginning this was very difficult because I was still learning vocabulary words. But I soon found the thesaurus tool in Microsoft Word, which quickly helped me find new words and their meanings. The more I wrote, the faster I became, until it seemed like the words just flowed from my head to my fingertips and I watched them form on the computer monitor. I wrote as much as my headaches allowed, thrilled to have found this new way to occupy myself during those sleepless nights. I was relieved to be able to describe my daily experiences and challenges and express thoughts I feared would upset Joan.

I often cried while I wrote because it stirred up all kinds of emotions, fears, and anxieties, but even if some demons were still too dark to release, I no longer felt so alone after purging the rest of them.

Every day I felt a little better as I learned the new technology that would enable Joan and me to build a new career together. I could think of no better way to spend the rest of our years as empty nesters than touring the country, giving speeches, and seeing places that would be all new to me.

Taylor was skeptical at first, based on the old Scott's distaste for traveling. "Oh, my God, are you serious?" she asked incredulously. "Are you honestly going to travel all the time?"

"Yeah, it would be great," I replied, wondering why she was so surprised.

Taylor informed me that I used to get short-tempered with any delay and minor obstacle in an airport or in the car. And some of that hadn't changed, Joan pointed out. Like the time I yelled at her to punch our location into the GPS when we took a wrong turn on a freeway interchange in Los Angeles.

"What am I supposed to put in, the 101?" Joan had asked sarcastically.

We ended up in the right place eventually, but it took some patience on everyone's part. The interconnecting maze of Los Angeles freeways could be maddening, but I was optimistic that things would be different for us in the future.

"Dad, you get frustrated with everything when you travel," Taylor said, still in disbelief.

"Not anymore," I said. "New Scott."

20

WITH MY PERMISSION, Lisa's husband tipped off the *East Valley Tribune* about my injury, and a reporter came to the house in August to interview me and Joan. Although it was getting easier to tell my story, I was still somewhat distracted and self-conscious. But it felt good once I got going, especially knowing I shared the same condition with someone else in the valley. Who knew, maybe the article would turn up some more of us.

We picked up several copies the morning it ran and also viewed the story online, where we were pleased to see a number of comments from well-wishers noting how devastated we must feel. As friends called and emailed with their support, I felt so revved up, I wanted to be a voice for everyone in the country who was struggling through life with a brain injury.

A few days later we came across a negative comment online from an anonymous reader who called me a liar and said I was faking amnesia for financial gain and to get out of lawsuits. I was shocked. Clearly this person had no idea how much pain my condition had caused me and my family. Our friend Johnna responded in our defense.

"Is this negative feedback going to happen a lot to us?" I asked Joan.

"I hope not, but people are going to think what they think and we can't stop them," she said, quickly guiding us off the website and on to more positive topics.

Curious about these lawsuits, I asked her about those as well. "Lawsuits are part of the aviation business," she replied. "We've been involved in several of them."

Seeing how much this negative comment had bothered me, Joan tried to protect me from any repeat occurrences by reporting this and any other nasty posts to the paper so they would be deleted, as this one was. But she couldn't be everywhere, and things soon took an unexpected turn for the worse.

Mattie pulled me aside after the next Brain Injury Association board meeting to say that she'd received a call from the same *Tribune* reporter. Apparently he'd gotten an anonymous letter claiming that I'd had a felony charge against me, that I was faking amnesia to get around a lawsuit, and essentially that I was a liar. I didn't even know how to react, so I just listened. From what Mattie described, the author sounded like the same anonymous critic from the newspaper's website, but I was stunned and embarrassed nonetheless.

Is this even true? It can't be. A felony? That sounds serious. I can't believe I could have done something like that. But even if I did, it couldn't be anything like rape or murder because Joan wouldn't have stuck by me. It would have to be some kind of other crime.

Mattie said the reporter asked if we vetted our board members and if she thought I was faking the amnesia.

Oh, my God, does this change the way Mattie thinks of me?

"Do you believe what this reporter says?" I asked.

Mattie assured me that she didn't and said she told the reporter that she couldn't talk about specific board members, but she didn't see why anyone would fake such a condition.

I apologized, saying I would certainly understand if she wanted to reconsider having me on the board, but she said that had never entered her mind. I was volunteering my time, she said; what did I have to gain by lying? She told me not to worry and to try to forget it. She'd only told me so I would know the reporter had called her.

After our conversation I was hurt that someone—even a reporter—would doubt me, but I was also relieved that Mattie trusted me and had chosen to stick by me. Still, I was frustrated and confused by the news, running over the conversation in my mind until I got home. When I walked in, Joan knew me well enough to read the angst on my face.

"What's wrong, honey?"

"Nothing," I said.

But Joan knew better. "Come on, did something happen?"

As I relayed my conversation with Mattie, I could see her temper rising. "I cannot believe he called Mattie and didn't call to ask our side. What mean person would call the executive director and question your integrity?"

"I don't know, but I just want to forget about it."

Joan, however, was not about to let this go. The next evening she told me she'd called the reporter from work, reminding him that we'd invited him into our home, even told him about Grant's heroin addiction. He replied that he was just doing his job, checking things out. When he asked her about the felony, she acknowledged that it was in the court records. We weren't trying to hide anything, she said, and would have explained if he'd asked us about it. Ultimately, he told her that he hadn't found any evidence I was lying or faking, so he wasn't going to pursue the matter further.

This being her first interview with a reporter, Joan told me that she never suspected the media would go digging into events from eight years earlier, looking for information that they would use to question my truthfulness about my amnesia. Nonetheless, I still felt protected by her and began to understand how she must have felt when I'd been in that role, protecting her, knowing that I wouldn't let someone hurt her without fighting back. Watching her, I was relearning the importance of sticking up for your loved ones.

But that didn't ease the pain of also learning that I'd been accused of breaking the law. I wasn't looking forward to hearing about it, but I did want to know the basic facts, so I asked her to explain. "Can you tell me about this felony charge?"

Leaning toward me from the ottoman in front of my chair, Joan took my hands in hers. Her eyes were drooping with compassion, and her lips were tight with the pain of having to deliver news she knew would be difficult for me to hear. "Are you sure you're ready for this?" she asked, giving me one last chance to change my mind. "Because I don't want to upset you."

I'd been driving myself crazy with the possibilities for the past hour, and I just wanted her to blurt it out and get it over with. "If I'm going to move forward in this new me, I'm going to have to know about the bad things that were part of my life as well," I said, bracing myself for the impact.

Reluctantly, Joan told me about the series of events leading up to the conviction eight years earlier that had haunted me ever since. I leaned back in my chair as she talked, feeling as if I were taking a new punch with every sentence: After having a falling out with a business partner and friend for whom I'd been managing jets, I withheld his aircraft's logbooks, hoping to force him to pay me what I felt he owed me, and his response was to file the lawsuit that Joan had tried to tell me about several months ago. He and I negotiated back and forth, and when he finally agreed to pay me a partial sum, I held out for past commissions and management fees as well. According to the paperwork I later found in my desk, I met with him and he gave me a check in exchange for the logbooks. But instead of walking away happy, I was arrested for theft, among other charges, by an undercover officer he'd brought along.

It was extraordinarily difficult for me to hear all of this, so much so that I glazed over as Joan was talking. I felt truly sick to my stomach. All I wanted to do was withdraw into myself to deal with the emotional heaviness I felt as I tried to process this development. It didn't help that Joan started crying as she watched me struggle with the news.

How embarrassing this must have been for Joan! I wonder what she thought of me when this happened. How could I have done something that must have disappointed her so much?

I knew what a felony arrest was from watching *Cops, The First 48, The Sopranos,* and *CSI Miami,* but I didn't understand much else of what Joan said, only that it sounded really bad.

"Why would I do that?" I asked.

Joan said I'd consulted with an attorney about the problems I'd been having with this partner, and although I'd told Joan some of what was going on, I'd largely acted on my own. Obviously, things hadn't worked out as I'd anticipated. Because we didn't have the money for a protracted court battle, I took a plea bargain in the criminal case and agreed to settle the civil case. At the time, Joan said, she and I felt our best option was to avoid going to costly trials in both cases.

All the criminal charges were dismissed but the theft, and even that was downgraded from a Class 2 to a Class 6 felony, which, she explained, was far less serious. Bottom line, she said, was that I served no jail time and got two years' probation, from which I was released a year early after doing one hundred and fifty hours of community service. The whole episode caused significant financial and emotional damage to my family, enough to cost us our house and force us into bankruptcy, but at least we were able to move on with our lives.

"I'm sorry that I put us in that position," I said. "It must have been difficult for you as well."

"It was horrible," Joan said, still crying. "What was most difficult was that you took things into your own hands—and I understand why; you were trying to protect me from the legal issues. If you'd told me your plan, I would have said no, no. But you made an error in judgment, and you've regretted it ever since."

"How did everyone else react?" I asked.

"No one else knows except you and me, and I think Mark may know, but I'm not sure," she said, explaining that we'd both wanted to keep this private because we saw nothing to gain from sharing it.

After we talked I found some court documents in my file cabinet and Googled some of the terms she'd used to try to better understand how this must have affected our lives together. Although I didn't have the emotional strength to read the paperwork carefully, I did notice that it said I had expressed remorse at the time.

Well, the remorse I'd felt then must have paled in comparison to the regret, embarrassment, and anger that I was feeling now.

What kind of man had I been to bring such devastation on my family? I must have caused pain to these other people involved as well.

I felt the utter stupidity of what I had done, but I also knew that I was going to have to be accountable and hold myself responsible for these events, even if I had no memory of them.

Later, as I was writing this chapter, I realized I was going to have to tell my family about this episode so they wouldn't learn about it for the first time in the book. And with that realization came the dread of having to reveal the shame and humiliation of the secret Joan and I had kept hidden for so long. How would I tell Taylor, who looked up to me so? How would I tell Grant, for whom I was trying to serve as a role model, and whom I was encouraging to clean up his own act? How would I tell my mother and father, who still saw me as their little boy and who were so proud of all of my accomplishments? And Joan's parents? I had no idea how they would react.

I resolved that the best tactic was simple honesty, to admit that I couldn't explain my past behavior—I couldn't even see how the old Scott could have gone there in the first place. If the same situation came up today, I knew history wouldn't repeat itself because I no longer acted alone on anything; I always consulted with Joan first.

I only hoped that my kids would appreciate the lesson that I'd taken away from all of this, that even adults make mistakes, and if we support each other as a family while we face such errors in judgment, we can overcome them and make our family even stronger. I was learning from my past poor decisions even now, using them as a guide on what *not* to do in the future.

As for my parents, I certainly didn't want to bring them any pain, but I hoped they would simply say, "So what. We know who you are, and this doesn't change that. We love you anyway."

. . .

By September, news of my accident had spread by word of mouth to my former coach and other alumni via my former teammate Phil, with whom I had spoken a couple times now since the accident. Joan still answered the phone, though, because I continued to be scared to talk to people I'd known in my previous life. When Coach Bill Mallory called, I wouldn't come to the phone, so he gave Joan a pep talk, offering us both words of support.

Joan had told me that I'd always been intimidated by my coach and held him in high regard, and perhaps I'd retained this emotion somewhere in my brain. The idea of having a conversation with him unnerved me, so I needed to know more about him before I felt I could even try. The articles I read described him as a strict and demanding disciplinarian, so when he called again some weeks later, I was surprised to find that he didn't fit this profile.

The compassionate retiree on the other end of the line gave me some heartfelt encouragement that was both touching and motivational. "I'm thinking a lot about you," he said. "You've always had a place in my heart, and I've always had great respect for you. You were always a hard worker and determined to succeed, and that determination is going to allow you to overcome this injury."

Some of the emails from my teammates, however, proved more difficult to process. I felt sad to read such kind words because I couldn't remember any of these men who had such strong memories of me and our time together on the field. Even though I appreciated the memories they shared, some brought me to tears.

Terry Clemans wrote that he'd been three years behind me in college and that I'd taken him under my wing. We were roommates for an away game in Lawrence, Kansas, during a brutally hot and humid Labor Day weekend in 1983. I was pissed, he said, when I saw trucks bringing in giant fans and blocks of ice to cool off the opposing team while we baked in the sun. But on top of that, it had rained like hell the night before so the Astroturf was holding water like a wet carpet. As we rolled around on it, we got more drenched with every play.

"I think that actually worked against them as it motivated you (and others) more than ever to kick their chilly Big 12 asses!" he wrote, referring to the Big 12 Conference, a league of college teams in the central United States.

Vince Scott, who said we'd played together all four years, wrote, "I can tell you, Scott, you were not always a man of many words . . . you were a man of actions. You showed your leadership by working hard in the weight room, off-season conditioning, and on the field. When you did speak, I guarantee you had everyone's undivided attention!"

But the email from Scott Kellar, who went on to play for the Indianapolis Colts, was the hardest for me because I felt no connection to the old Scott he described. Reassuring me that of anyone, I would be able to overcome this injury, he said we met when I was a junior, with the promise to become "a devastating offensive lineman."

You never knew it, but every time I had to go one-on-one with you, I was so nervous I would feel sick. You had one gear, and that was 100% all of the time. You played with what Coach Mallory always referred to as "a prick in your vein." You came off the ball with a nastiness about you. . . . You were the best offensive tackle I ever played against in college. I credit you for helping me get to the NFL. It was an honor to play with you and, yes, get my butt kicked from time to time.

What was so very impressive was the presence you exuded whenever you walked into a meeting room, locker room, and the field. You were the consummate leader, and I, as well as many of us, had and still have the highest respect for you. However, more important was the fact that you were and are a great person. You never looked down on anyone. . . . Even when I was a rookie, you rolled me off the line of scrimmage then proceeded to slap my back and told me to keep working hard. You were always just a great guy to be around. You were a great role model and someone I looked up to.

Scott, I hope for nothing but the best for you. I know you are going to come out of this better than you ever imagined. Keep striving every day.

Reading this, I wanted to feel like the old Scott again and have the same kind of positive impact I'd had in college. Given my recent discoveries about my other past behavior, I wondered how the old Scott could have had such polar opposite impacts on the people in his life.

Despite the bittersweet nature of these emails, I always thanked the sender and asked for more stories because they made me proud of who I'd been, what I'd stood for, and how I'd dealt with adversity and success. I now had the chance to pick and choose what characteristics I wanted to keep or discard as I transformed into the man I wanted to become.

These messages inspired me to remember that the leadership I showed at NIU was based on simple morals and values. Thinking perhaps that my legal problems had come about because money issues had blurred my integrity, I vowed that when making decisions in the future, I would adhere to the pure, grounded, and untainted values of my NIU days.

. . .

Over the next several weeks and months, the story of my accident and recovery caught on with a wider audience as Joan and I were interviewed by other newspapers, radio shows, and local television news stations. Soon a producer from ABC's *20/20* called to tell us they were interested, then called back a few weeks later to say

Nightline was going to interview us instead because its producers could air my story much sooner.

In the meantime, I followed up on my doctors' recommendations to see a therapist, and I met with Dr. Philip Barry, a neuropsychologist, to help me improve my coping skills. I didn't see this as a long-term project, but he told me progress wasn't going to happen overnight. This, of course, was not what I wanted to hear. I felt like I wanted—and needed—to get better now, as in right now.

Dr. Barry made me realize that life is full of challenges, and just because my previous life had been deleted didn't mean I wouldn't face some of the same obstacles as before. He also gave me some helpful advice: it was okay to be scared and frustrated, and if people didn't understand what I was going through, to hell with them. Let the new memories come, grab them, and make the most of them. So that's what I set out to do.

After just two sessions with him, I felt like I had things under control so I stopped going. Looking back later, I realized that this, sadly, was nothing more than wishful thinking on my part. Other than Joan, I still had no one to vent to, so I kept the things I couldn't share with her locked up inside, and they began to fester. I had thought I was strong enough to figure out how to deal with my daily challenges all by myself, but I was mistaken. Frankly, I'm not sure anyone is.

21

G RANT GOT KICKED OUT OF the Donna's House program
for using drugs six weeks after arriving there in late June,
and we later learned that he'd been using all along. There
went another $1,600 down the tubes. Out of options in southern
California, we brought him back to Arizona and into the house,
where we thought we could keep an eye on him. But, as I'd been
learning, addicts can be sneaky and cunning.

At times I found myself looking at him, searching for some kind
of connection between us so I could try to understand why he was
making the choices he did, who he was, and how he felt inside.
Maybe if I understood him better, I could find a way to help him
more than I had been able to up to this point.

I could see from family photos that Grant had resembled me
more when he was younger, but only until he hit sixteen. Today
he didn't look much like Joan either, except for his deep blue eyes.

I didn't understand why Grant dressed the way he did or had
mutilated himself by stretching giant holes in his earlobes with
rings called gauges. This seemed so disrespectful to his mother and
me, who had brought him into this world in one perfect piece. So,
after Grant's relapses that summer, I told him so. "I don't want you
wearing your pants so low that I can see your underwear," I said. "I
don't want to see your piercings or your gauges in, either."

Grant didn't fight me on the pants, but he refused to remove the gauges. "I can't take them out because they'll close," he said.

Our strongest commonality was our athleticism, but even that was in the past. Joan told me that back in his motocross days, Grant seemed to have little concern for his body's safety, probably the same indifference I'd had to play football and get smacked around for as many years as I did.

Could I have taught that to him in some way? And does that have something to do with the self-destructiveness he is now wreaking on his body with drugs?

Grant had risen to the top in hockey and motocross because of his natural talent but had lost that edge once his teammates started to work harder to make up the difference. Joan said the difference between us was that, when faced with a challenge or hard work, Grant lost interest whereas the old Scott had dug his heels in even harder.

I also struggled to understand why he felt the need to cover himself with tattoos. I didn't have anything against them per se. I'd actually discovered in the hospital that I had one myself. Not knowing what it was, I tried to rub it off—hard, but it didn't budge, and I didn't mention it to anyone for fear of looking stupid.

Once I started watching TV at home, I started seeing other people with these colored markings and also learned that the five-by-five-inch design near my right shoulder was an American flag, an eagle, and the letters *USA*.

"What do you think of my tattoo?" I asked Joan.

"I don't like tattoos, but it's something you wanted, and it's your body," she said. "But it's not a shameful tattoo. It's a nice tattoo, and it's covered."

I figured I must have had a good reason for getting it in the first place. I'd already asked Joan how I'd gotten all the scars on my body, and she said most of them were from surgery to fix injuries. But getting a tattoo was a choice.

"Why did I get this?" I asked.

Joan explained that I'd gotten it in 2002—after September 11, 2001, which she and everyone else called 9/11—because I'd always been patriotic and had a tremendous respect for people who served in the military. The eagle and the flag represent courage, dignity, freedom, and my pride in my country.

The tattoo was something Grant and I had in common; I just felt he took his interest to an extreme. So what, I wondered, did Grant's tattoos mean to him? I'd seen the two on the inside of his wrists—*Truly* on the right and *Blessed* on the left—after I got out of the hospital. But the first time I saw the one over his heart with Taryn's name in black cursive letters was that summer, when he took off his shirt to go swimming. It put me off, and I hoped he had a good reason for getting it.

"Why do you have Taryn's name on your chest?" I asked.

"To keep her close to my heart," he said.

"Are you still happy that you got that tattoo, or do you wish that you'd never gotten it?"

"Are *you* happy that I got this tattoo?"

"No, honestly, I would prefer if it wasn't over your heart. She has a very special place in my heart and your mother's heart, and I don't think that she's in *your* heart."

I wasn't trying to be mean, but he hadn't even been born when she died, and he just didn't seem like the spiritual type. I also didn't think that he felt or understood the true meaning behind his words. When Taylor ultimately turned eighteen, she too got a *Taryn* tattoo, but it was a delicate one with angel's wings, hidden at the base of her neck. That didn't bother me at all because she *was* more of a spiritual person.

"Why do you like tattoos?" I asked.

He said he saw himself as a free spirit, a rebellious nonconformist, and wanted to be a body piercing artist, but all I could think was that he couldn't be more different from me, who, from what I'd heard, had always wanted to be normal and even now just wanted to fit in. I couldn't see how Joan and I had raised a son who was so outside the norm, especially when we'd also raised a daughter who was so much more like us.

Taylor had been helpful in giving me some insight into Grant's and her generation, but even she said she didn't understand how her brother turned out so different from her.

He had other tattoos I didn't like either, such as the eight-inch bold dark letters spelling *relentless* that went from his armpit to his waist, which Joan said he'd gotten as soon as he turned eighteen. I had a hard time dealing with this one too because I didn't understand it. If I had a thousand words to describe Grant, *relentless* wouldn't be one of them. If I had to pick a word, it would be *lost,* same as me.

Joan and I had talked quite a bit about the parallels of Grant's addiction and my memory loss, which had forced each of us to search in his own way for his true identity and who he was as a man. Grant had lost his sense of purpose and his direction; he'd lost his way. My brain injury had caused me to lose my sense of self and caused my confusion. But I saw a potential for connection between us here, as each of us tried to fight through our own emotional pain, so I tried to use these shared struggles to bridge the gap between us.

"If anyone can understand how you feel, Grant, it's me," I said. "We've both been through a traumatic brain injury. I understand what it's like to be lost."

After my own experience with depression and getting some relief from Cymbalta, I also tried to talk to him about the depression I'd seen in him and the benefits of antidepressants. "I know what it's like to not be able to get out of bed, to feel hopeless," I said.

Grant had tried the meds on and off over the past few years, but he'd always stopped taking them after a month or so. His excuses ranged from not liking the stigma to not having the money to buy the medication, and his current rehab wouldn't let him take any medications. Of course, when he did have money, he'd turn around and spend it on recreational drugs instead.

In the end, my attempts to have these heart-to-heart talks just didn't seem to do any good. Apparently Grant wasn't able to connect with me on that level. He seemed to resent me, and that left me frustrated, disappointed, and sad.

. . .

It wasn't long after he'd moved back in with us that I noticed about ten days' worth of my OxyContin pills were missing from the bottle, which I'd hidden in my sock drawer. When Joan and I confronted him, he said, "I know you're not going to believe me, but I didn't take them." I had no proof, but it wasn't hard to do the math. I took the medicine three times a day, and thirty tablets were missing.

Perhaps his conscience got the best of him, but several weeks later he admitted to the theft and apologized. We felt it was a big step for him to come clean like that, so we gave him a final warning.

"This is your last chance," I told him.

From then on, we hid my OxyContin in the house safe, which was in my office closet and required a combination and a key to open. Grant knew the combination because we stored our trust documents and wills in there in case something happened to us. However, now that he was home, we thought it best to hide the key.

I never thought he would find the key, which I'd hidden in an organizer on my desk, but that he did, several days before Joan and Taylor were due to go to Hawaii with Joan's mother. When I saw that at least fifty of my pills were gone, leaving about one more day's worth, Joan and I confronted Grant once again, and he denied, denied, denied.

"I know what was there," I said. "There's no way for this number of pills to be gone without someone taking it, and the only logical choice is you. There's no way I took an extra fifty pills by mistake."

But Grant stood firm. "There's no way I can get into the safe because you need a key and I don't even know where the key is," he said. "I haven't been in your office since you told me not to go in there."

I knew he was lying, but I was so fed up with him I just wanted to get away from him. While Joan and Taylor were in Hawaii, I planned to relax on the boat for four days to clear my head and start honing message points for my new speaking career. After recently losing a refill prescription for the pain meds, I felt it wasn't an option to ask my doctor for another—or to confess that my son had taken my supply. Although I'd been taking the OxyContin regularly, I was down to a pretty low dose, so Joan and I figured it wouldn't be a problem to go off it and that I could try to deal with the pain, using some backup Percocet I still had if necessary.

I had mixed emotions about Joan's departure. I wanted her to go and enjoy herself, because God knows she needed a break from me and certainly from Grant, but I knew it would be difficult without her. Before they left I made sure they had their boarding passes and hotel confirmations; I even went to the bank to get them some cash.

After dropping them at the airport, I started my six-hour trip to Oceanside, figuring I would arrive about the same time they landed in Maui. While we were gone, Anthony agreed to stay at our house to "babysit" Grant and make sure he didn't slip out to buy drugs.

Shortly after I got to Oceanside, Joan called to say they'd landed and were on their way to the resort, but she was more worried about me because I wasn't feeling too hot. She suggested that I go to urgent care in California for more OxyContin if I was so worried about Dr. Lanier's reaction, but I told her I'd just tough it out.

On my second day there, I felt like I was coming down with something. My body ached and I couldn't sit still, but it hurt to move. I was either too hot or too cold, I couldn't concentrate, and I felt disoriented. This was my first time being sick since the accident, and Joan thought maybe I had caught the flu due to the stress of the past ten months.

After the third day of feeling crummy, Joan and I decided I should drive home and let Dr. Lanier look me over. Still suffering from insomnia, I headed out around 3:00 A.M.

Seeing no choice but to be honest, I explained to the doctor that I hadn't taken any OxyContin for four days because Grant had stolen my pills, and she immediately knew what was wrong.

"You're not suffering from the flu," she said. "You're going through withdrawal. It's not a good idea to just stop like that."

Asked how long the detox would last, she said, "You should feel better in a couple of days."

"That's fine," I said, "but I'm going to go off the drugs entirely and see how the headaches are. Maybe, just maybe, they'll go away."

Grant was still asleep when I got home, so I thanked Anthony for watching the house and told him he could go. Now that I knew what was wrong with me, I was angry that these four days of feeling ill had been forced upon me by my own son. The situation clearly called for a verbal kick in the ass, so I went into Joan's office, where he was sleeping on the spare bed, and kicked him in the leg.

"Get the hell up," I commanded. "I want to talk to you. Now."

Startled, Grant muttered, "What's wrong? Why are you here?"

"Never mind," I replied. "Just get up."

He joined me in the living room several minutes later, groggy, with his hair tousled. When I told him I'd just come from the doctor's office after feeling like crap with the flu for the past four days, he acted concerned.

"That sucks," he said. "Are you okay now?"

"Well," I said, "I found out that I don't have the flu, Grant, but that I'm going through withdrawal. And do you know why? Because you stole my medicine."

"Dad," he said quickly, "I didn't take your pain pills. I can't even get in the safe."

"I think you're a liar, and I'm so pissed off I don't even know what to do with you right now." After pausing to let that sink in, I went on. "From here on out you're not going to lie around this house anymore. I want you working, cutting grass, cleaning the garage, and whatever else I tell you to do. Is that clear?"

"Okay, whatever," he said with a mix of apathy and anger.

While he was mowing the grass, I called Joan and told her what the doctor said. I told her to have fun for the rest of her trip; I was going to be working our son from morning till night for the next couple of days.

"We're going to have to make some decisions about Grant when you get back," I said.

Joan agreed, saying he'd been mouthy with her the last time she'd talked to him on the phone from Hawaii. "He's using again," she said. "I know it."

. . .

After Joan and Taylor flew in, Taylor headed over to Anthony's to avoid the scene with Grant she knew was coming.

Still maintaining that he wasn't using, Grant had persuaded us to let him go to Las Vegas with his girlfriend for the weekend because he'd told us she didn't use drugs, so Joan and I planned to talk to him peacefully when he got back.

That changed, however, after Joan cleaned the bathroom, where she found several pieces of a gray cottony substance, a small black nugget of what she presumed was black tar heroin, and the tip of an insulin syringe. Joan then searched her office and the bed Grant had been sleeping in, finding more cotton and some wadded-up squares cut from a plastic shopping bag.

"Look what I found," she said, so upset she was shaking.

The syringe was the only one of these items that Joan had seen in real life, but she'd seen the others online and on TV. I, on the other hand, was clueless. "Where did you find these, and what are they?" I asked.

She tried to explain what little she knew about how Grant procured and used his drugs, saying the cotton was to filter out impurities when he shot up.

"This is what the heroin comes in," she said, pointing to the plastic squares. "He's using in our house."

Grant had obviously crossed the line. "We can't have this in our house," I said, horrified by the implications.

"He's got to leave," she said, just as mortified as me.

When he came through the front door two nights later, we asked him to meet us in Joan's office, where I stood with my arms crossed and Joan sat at her desk.

"Is there anything you want to tell us?" I asked.

Grant, who we later learned was still high from using that morning, was in smart-ass mode. "No," he said. "About my trip?"

"No, about the drugs that your mother found while cleaning the bathroom."

Grant paused for a moment, then copped to it. "Yeah, I'm not going to lie. I used drugs in the house," he said.

"In our *house*, Grant?" Joan repeated incredulously. "We can't trust you, and you aren't allowed to stay here anymore. You need to get help."

"That's okay," Grant said, adding nonchalantly that his girl-friend would let him stay with her family until she got her own apartment.

His sarcastic attitude only made me angrier. "You are the stupidest kid I know," I said. "We've spent nearly $75,000 dealing with your drug addiction, rehab, detox, and medical appointments. We've tried to help you, and you refuse to follow our rules in this house. I don't care if you become homeless. You need to leave and get your life in order."

"Fine," he said.

"Don't you even care that you're hurting us?" Joan asked.

"No, I really don't," he replied, looking right at her with such insolence and disrespect that I was furious on her behalf.

Where did he think he got off talking to his mother like that?

I grabbed him by the neck and pushed him out of the bedroom and across the next room until I had him against the open door, which was flat against the wall. "It's time for you to leave *my* family alone. Get the hell out of here!" I yelled. "And don't call us."

Grant just laughed, which only made me more furious. It really did seem like he could care less. "Say good-bye," he said as he opened the front door. "This is the last time you're ever going to see me. I have enough heroin in my sock to overdose."

Stunned by his last remark, my anger dropped a few notches. "I can't stop you from doing that, but we love you," I replied just before he shut the door.

By now Joan was crying hysterically. Thinking logically, I figured his overdose threats were empty and manipulative and that he would simply find a ride to his girlfriend's house. But Joan was clearly caught up in her emotions, fearing that he might just follow through.

We had another rough night, with Joan worrying where he was, but she calmed down after she got hold of his girlfriend the next day, who said she'd gotten Grant a hotel room for a couple of nights. When he called to tell Joan that he was safe, she asked, "Are you going to hurt yourself?" and he said no.

Grant persuaded his girlfriend and her family to let him stay with them. We weren't happy, but we were relieved that at least we knew where he was sleeping, and it wasn't a park bench.

· · ·

Joan's hysterical crying was just one sign that she wasn't handling things as well as usual. After being so strong for so long since my accident, she'd been showing increasing signs of stress and had been growing frustrated and angry with me more easily than before. These days, an offhand comment from me would send her into a tirade.

Since the accident I'd gradually been joining in more of the joking banter our family did with each other. Once I'd gotten past my sensitive stage, I'd learned this was all done in a good-natured, playful way. But sometimes lately I'd noticed that when I tried to joke with Joan, she would go off on me in a bitchy tone. Honestly, it was like waking up to two different people. One day she was nice and sweet, and the next she'd be so difficult I'd want to run away from her. Concerned that I was doing something wrong, I tried to lie low for a while, but I started taking the whole thing personally, attributing her moodiness to my inability to be a productive husband and provider for our family.

Is Joan starting to resent me for not being the man she once knew and married? Am I ever going to be that guy again?

Then I started noticing that she was also directing her irritability at Taylor, yelling at her for minor infractions such as leaving a knife covered with peanut butter on the counter. If she was screaming at our sweet and well-behaved daughter, I figured she must be under a tremendous strain.

When Joan and I talked about her mood swings and erratic behavior, she summed them up as PMS. Once I had a label to place on it, my first thought was, *I've heard of this, but how do you get rid of it?*

Once again, TV helped me out. I often watched *Everybody Loves Raymond* to gain a better understanding of how families interacted, and one episode seemed particularly relevant to our current dynamic. Raymond's wife, Debra, had a bad case of PMS, so he bought her some over-the-counter pills and read to her from the box about the symptoms it was supposed to cure, including bloating, headaches, fatigue, and muscle aches.

"There's nothing in here for bitchy," Debra noted sarcastically.

"You probably need a prescription for bitchy," he replied.

I felt like Raymond was speaking directly to my situation. It was kind of a relief, actually. Here I'd thought it was me and our household dramas, but it was simply the three most dreaded letters in the male language: *PMS.* Now that I knew what it was, I began to notice it started right before and continued during her menstrual period.

After I'd started taking the Cymbalta, I'd felt the changes in my own mood. And after seeing the TV commercial for the drug, I wondered if maybe Joan was depressed too, grieving the loss of the husband she used to know. So, very cautiously, I broached the subject with her.

"Maybe Dr. Lanier can help you with some Cymbalta," I said. "I've found it's been helping me cope with the stresses of life. Maybe you would benefit from it as well."

Joan mentioned that she'd already tried some antidepressants for PMS in the past and hadn't seen much improvement, but she said she'd talk to Dr. Lanier. In July Joan began taking the medication.

Within a few weeks Joan's moods evened out somewhat and the time around her next period was much improved. But even then, some other thorny issues presented themselves.

After the news stories started coming out, I sensed that Joan was becoming increasingly jealous of the instant attention I was getting, of the opportunities to tell my story, and of my decision to launch a new speaking career, which was a longtime career goal for her, not to mention the repeated suggestions that I write a book.

Confused and frustrated, I was worried that we were headed for trouble if this continued. As much as I tried to include Joan in my media interviews, this gesture never seemed to fulfill her needs. I thought I was lost before in life, but there is no worse feeling than having a simmering uncertainty about what your wife wants and a nagging inability to figure out how to satisfy her.

22

W ITH MY FIRST HALLOWEEN coming up, Joan and I committed to take part in a charity event put on by Save the Family Foundation of Arizona to benefit underprivileged children, and we were encouraged to bring Taylor and my nephews Noah and Aden along.

I was all for helping kids, but even after watching Halloween episodes of my regular shows on television and seeing the costumes and candy at Target, I still didn't get the idea behind this strange ritual, so Joan tried to explain it to me. "It's a holiday that allows you to be goofy, and you give kids candy," she said.

To me the notion was bizarre, but I generally enjoyed doing anything that Noah and Aden enjoyed, and I figured I would comprehend it better once I saw it. No longer the old Scott, who questioned Joan's interest in participating in charity events, I wanted to be there *with* her, doing something to help. As an NFL alumni representative, I was going to meet the organizers and observe the festivities firsthand because my group was considering whether to support this charity in future events.

We met up at the venue, the Hilton hotel near the Phoenix airport. While Anthony and I waited outside the ladies' room, Joan and Taylor had the boys change into the costumes that my sister Bonnie had sent from Chicago. Three-year-old Aden was cold, so he pulled his red and gray nylon Transformers outfit over his clothes and put on his pointy-eared Batmanesque mask that covered his face from the nose up. Six-year-old Noah was a Ninja Warrior, wearing a plastic six-pack over his stomach and a stretchy hood mask with an opening for his glasses.

When the boys emerged they were beaming with pride, which made me feel excited right along with them.

"Awesome!" I said. "You guys look great!"

The event started in a room of parents and their three hundred children, who were loaded up with pizza and party food, including designer cupcakes as well as the homemade kind with smeared frosting. Considering these families had recently been homeless and were now trying to get their lives back on track, I figured this was probably the first time all year that they'd been indulged with goodies like this.

For those few hours we had a blast. In that hotel all their troubles—and mine too—were forgotten, receding into the laughter and the bubbly silliness of children having fun. It made me feel like a kid too, so this was an effective way for me to experience what it had been like for me *and* my kids, growing up. When Joan noted that these children seemed to prefer the homemade cupcakes over the fancy ones, I wondered if, like me, they stuck with what they knew because, with all the turmoil in their lives, they needed the comfort of the familiar.

One surprising thing I learned was that homeless people weren't just the drug addicts or the older men with straggly beards I'd seen on TV, a group I feared that Grant might soon join. I was surprised and saddened to see so many mothers with children in this group, a segment of society I'd never pictured having to live in a car or under bridges until the organizers told me otherwise.

As we took the kids on a trick-or-treat tour of the first floor, where every room had been decorated by hotel employees, I was impressed by the kids' creative ploys to get candy.

"Show us a trick," a staff member told a little girl in a princess costume, who danced and pranced around to earn her treat.

Aden refused to go into any of the rooms with haunted-house themes because he was too scared.

"Don't you want to go inside and see?" I asked. "I'll go with you."

"Uh-uh," he said, stiffening his arm against the door frame to keep anyone from pushing him inside.

"It's just people with costumes," Joan said reassuringly, but he still wouldn't budge, happy to let the staff bring him candy from inside. The fearless Noah needed no encouraging, however, plowing right in and exploring.

Melia Patria, the *Nightline* producer, followed me as I vicariously experienced the kids and their glee. She'd been trailing me with a camera for several days, from 9:00 A.M. until 9:00 P.M., sometimes as late as 11:00 P.M., asking me for one more story before we turned in.

"Tell me about that," she'd say.

During the four days Melia was with us, we pulled out the wedding footage that my sister Bonnie had found in her garage and let Melia capture my reaction as I watched it for the first time with Joan on the couch. It was grainy, only a couple of minutes long, and had no sound. It showed us walking out of the church and getting into the limo as people threw rice on us. I'd wanted to know how I'd felt that day, and the video answered that question: grinning ear to ear, I looked like the happiest guy on the planet. I only wished there was more footage to watch.

When Melia was done filming, she told us she would return soon with Bob Woodruff, the ABC reporter who had suffered a severe brain injury when hit by an IED in Iraq. This was quite an honor for me and Joan, because we'd seen him on television and read his book. We were pleased that he wanted to cover our story, assuming that he would be more compassionate after suffering a brain injury himself.

. . .

Even though no one but Joan wanted Grant to join us for Thanksgiving dinner, he came over and acted his usual passive-aggressive self. We captured the event with the TV camera that Melia had left with us to film any firsts for me, such as handing out candy on Halloween or stuffing the turkey.

Earlier that week Joan had asked me to do her a favor on the day after the holiday, which she explained was known as Black Friday.

"Honey, would you go to the store for me on Friday?" she asked. "It's a big sale day. This is a little different from what you usually do, but you'd really help me out because I can't be two places at once."

"Whatever you want me to do, I will," I said.

Joan wanted me to buy Taylor a sewing machine for college. I'd have to get up early to be there by 5:30 A.M., when the store opened, which I said was no problem.

While Joan and Taylor were still at the mall taking advantage of the clothing sales from midnight to 5:00 A.M., I left the house for Joann Fabrics. Aiming to arrive early and be at the front of the line, I was shocked to see two hundred crazed women already queued up, coupons in hand, discussing what they were going to buy. I was damn near afraid for my life; it seemed like these women would plow me over to get what they wanted.

"Why are you here?" asked one of less threatening ones.

When I told her about the errand my wife had sent me on, she laughed. "What a good husband, coming here and fighting this chaos," she said. "I think your wife pulled a fast one on you, because you have no idea what you are about to see."

The line started moving as soon as the front doors opened, and I felt like I was in a pack of hyenas chasing a rabbit. Once we got inside, pointy elbows were flying as the women competed for pieces of fabrics and silly Christmas decorations. I ran the other way as fast as I could, looking for an employee to direct me.

"Where are the sewing machines?" I asked.

"Follow me, sir," she said.

Doing as instructed, I found the Brother electronic sewing machine, with all the bells and whistles, that Joan had circled in the advertisement. I grabbed the box and headed for the register with a coupon in my pocket for an additional 10 percent off. I didn't get very far, though, because a masculine-looking drill sergeant wannabe, whose jeans were too tight for her oversized frame, stopped me in my tracks. "You can't leave this area," she said. "You have to pay for that here."

"Okay," I said sheepishly.

As she rang me up, I handed her the coupon. "You can't use that with a sewing machine purchase," she scolded, clearly frustrated with me.

"My wife said I had to use it, and I better use it," I replied.

"I'm sorry, but tell your wife it's not good on electric machines. It says so right on the coupon."

With that, I paid and got the hell out of there, happy to have escaped alive. As soon as I got outside I called Joan, hoping against hope that the news helicopter hovering above me wouldn't catch my humiliated face on video as I stood outside a fabric store with a sewing machine under my arm. Joan answered on speaker phone, and I could hear her laughing.

"How was Joann Fabrics?" she asked. "Did you get it? Did you have fun?"

By then I was able to find the humor in the situation. "I think you took advantage of me by feeding me to the wolves," I said.

She laughed again, as if she'd known exactly what she'd set me up for.

I wasn't sure if Joan pulled a fast one on me, but in the future I would be more savvy and honor her Black Friday shopping requests only if they were for manly places, such as electronic stores, where Joan said I used to go.

. . .

Gearing up to meet Bob Woodruff was nerve-racking, but the anxiety soon faded once we started sharing stories about our brain injuries and memory losses, as if we were comparing battle wounds. Even though his were from an actual war, I felt like my body and my mind had been waging their own battle.

Bob explained that his pain had been much more physical than emotional, but he too had suffered headaches for a long time and lost part of his sight. Joan and I had read about his pain, how he'd dealt with his new disabilities, and the difficulties he'd faced returning to the career he'd always loved. Although I couldn't remember running an aviation company and didn't know if I would ever have a desire to do so again, I hoped that he would give me some insight into how he'd managed to persevere.

As we talked it soon became apparent that, like me, the love of his family was what had gotten him through the nightmarish recovery we'd both experienced. I noted that, also like me, Bob had tried to keep a positive attitude and to work through the pain to reach his goals.

While the crew was setting up to shoot some footage of us walking toward the fountain in the backyard, Bob and I shared a few private moments, talking man to man. He said he had retained his wealth of knowledge, but he sometimes had trouble remembering the right words to say. I'd lost all my knowledge, I told him, and had a similar word retrieval problem.

"I don't know if I ever would have made a full recovery if I didn't have my memories to fall back on," he said. "That's what really helped me to focus on getting back to who I was. I wouldn't want to trade places with you for anything."

His last words paralyzed me as I felt the shock of their impact creep through my skin.

How bad off am I, really, that this guy, who had a big part of his head blown off by a roadside bomb and has gone through so much physical pain, wouldn't want to be me?

I knew he didn't mean to be negative or hurtful. I was sure he only wanted me to understand how far he'd come in his own recovery. As wounded as I felt, I tried to put his comment aside and focus on his drive and determination, which were truly inspirational to me. I too was on a quest to make it back and be successful once again in whatever I chose to do. If I spent too much more time thinking about it, Bob's words would have set me back months in my recovery. But at the time, it felt like they already had.

. . .

It had been five months since we'd celebrated my forty-seventh birthday, but Joan and Taylor thought it would be amusing to get me another birthday cake to commemorate the first anniversary of my accident on December 17. I guess I couldn't blame them. Whenever I didn't know something or made a mistake, I'd throw out my "get out of jail free" motto: "Give me a break; I'm not even one year old yet."

That night after dinner they turned out the lights and brought out a chocolate cake, topped with a flaming #1 candle and Happy 1st Birthday spelled out in red icing. I had to laugh at the gesture.

"You guys are funny," I conceded.

They sang "Happy Birthday" to me, and even though I was now an old pro at it, I wasn't about to sing it to myself.

"Does it seem like a year?" I asked.

"When you don't know something or when it's something we all know you knew, it feels like it just happened," Taylor said. "But other days, when nothing comes up, it seems like a long time ago."

We debated the differences between the old and new Scott, my fresh feelings of shock about society, including how people treated others with such hostility and anger and how much and how fast some people ate.

I'd had similar conversations with Mark, who told me I was much more mellow but much less self-confident and assertive since the accident. Before, he said, "you never took no for an answer. No matter what, you'd find a way to accomplish it. That's gone."

I didn't know if I'd really felt that way before, but I certainly didn't feel I possessed those personality traits now. What I did know was that the more knowledge I gained, the more confident I felt. Meanwhile, there were a whole lot of things I still didn't know and struggled to understand, including concepts the old Scott once comprehended.

Joan said I wasn't a "creative thinker" anymore. Before, I was more jaded and bitter from the hard knocks the business world had thrown at me, but I was good at coming up with innovative marketing ideas, such as changing our marketing focus to the jet debit card—what Joan called "thinking out of the box."

Today, I felt like I lived *in* a box, and yet it felt comfortable in there. Following a routine and doing the same things over and over helped me cope with daily life, with its constant bombardment of new information and countless choices. For example, I'd liked eating buffalo wings before, but I ate them frequently now because they were a comfort to me—something safe I *knew* I liked.

Today, I also needed people to show or prove things to me in concrete terms before I could understand or believe them. Intangibles and abstract concepts that I couldn't see, touch, or smell made me agitated because I couldn't get my head around them.

Take my brain injury and the SPECT scan, for instance. Once I saw the images of my brain scan on paper, I began to understand the concept of reduced blood flow to those areas. But that understanding was crystallized when I saw Dr. Korn explaining the test to Bob Woodruff on *Nightline*—and showing him the corresponding orange and blue parts of my brain as compared to a normal brain on the computer screen. Without seeing those test results, I would probably still be wondering.

They say that addicts and alcoholics are black-and-white thinkers too, so that gave me something else in common with Grant. "I can surrender to the idea that I'm powerless, but I don't know if I can surrender to a higher power, and I don't know what a higher power might be for me," Grant said to explain his problems with the twelve-step program in AA. Given my problems understanding religion, I could see why he was having such trouble.

Joan had told me that she was still a believer, after being brought up in a religious Lutheran family, and although her parents still went to church and Bible study every week, she stopped going because she had problems with "organized religion." Sometimes I took her word for things, and the idea that there is a God was one

of them. But that didn't mean I wasn't still struggling to understand the concept.

*Who's taking care of Taryn if there's no God and no heaven? But
if no one can show this person and this place to me, how can I believe
in them?*

"Where is he?" I asked Joan.

"He's everywhere," she said.

"How do you know this?"

"You just have to believe," she replied. "You don't have to see to
believe, but if you look at the mountains, the oceans, the human
body, how can this be created by anything other than a God?"

My hope, of course, was that something or someone was in fact
taking care of Taryn and that she was someplace peaceful and
beautiful, not just buried beneath the headstone I'd seen in our
scrapbook. Still, the more questions I asked about religion, the
more questions I had.

When we go to Chicago in the spring, Joan says she will take me
to the church she'd attended growing up, where we'd also gotten
married. Knowing I'd been brought up Catholic, I told her that I
wanted to go to one of those services too. It was all in the name of
relearning and rediscovering, and I still had plenty of that to do.

. . .

When people asked me, "What is it like not having memories of
your past?" I tried to relate my explanation to a common experience. I usually asked if they'd seen the movie *Family Man,* starring
Nicholas Cage as Jack Campbell and Tea Leoni as his wife, Kate.

This had been a family favorite before my accident, and although I'd probably watched parts of it fifteen times since, it was
still difficult for me to get through because the plot had so many
parallels to my situation. I shared many of the frustrations and
feelings of loss that plagued Jack, a successful and materialistic
Wall Street deal maker who wakes up Christmas Eve in the life he
would've had if he'd married his college sweetheart. He suddenly
has two kids, a dog, a career as a salesman at a suburban tire sales

outfit, and an attorney wife who helps the poor, but no memory of how he got there.

The look of confusion on Jack's face when he wakes up was uncannily familiar. I'd felt the same way when I arrived home from the hospital, as if I was having a bad dream. We'd both married our college sweethearts, were both successful businessmen, and had both discovered that our lives had completely changed, but we didn't know why or how or who we were.

It was hard to watch the fear in Jack's eyes when he realizes his new life is not going to change because that was also true in my life, and I was reminded of it every day.

People have told Joan and me that our story is like a movie. That may be true, but for us it's our reality, and every day brings us a new obstacle to overcome. Still, like the new Jack Campbell, I am and always will be a family man. And like Jack's family, ours always tries to joke and laugh during the toughest times because sometimes it's the only way that I can deal with things and stay positive. Apparently this is one thing that has changed for the better since my accident. Before, I'd get pissed when I made mistakes, unable to laugh at myself. Now, I found the innocence of my mistakes comical, and if I didn't laugh, I'd probably just cry.

I had to admit that, as my reeducation continued, some lessons were still more entertaining to others than they were to me. Like the afternoon that Joan asked me to cut an onion but forgot to warn me about the ramifications. There I was, slicing away, when my eyes started burning like hell and tears came pouring out.

"What is this?" I asked, startled and confused.

Joan said this was a normal reaction, then chuckled at my culinary misfortune.

"What kind of seeing-eye dog are you?" I quipped.

. . .

My second year of yuletide cheer went much more smoothly. With Taylor's help, I bought Joan some Victoria Secret mango lotion and a massage gift certificate, and she got me a book titled *100 Days in Photographs: Pivotal Events That Changed the World,* along with some other thoughtful yet inexpensive gifts.

Seeing that my brain had been healing and I'd taken so much pain medication during the previous Christmas, I didn't remember much about the ornaments or where they went on the tree, so as we hung them I asked Joan and Taylor to tell me the stories again. At first I was concerned about why I'd forgotten something so important *after* my accident, but I soon got over it. It was good to hear the stories once more—including the ones about the precious Taryn angels—without the trauma of having to learn about her death all over again.

Joan put on some silly Christmas music, and I had to tease her as she and Taylor sang along. "You can remember the words to these, but you can't remember where you put your car keys?" I joked.

She smiled and pushed me away playfully, saying, "You don't understand how many times we've heard these songs."

I put up the decorations and lights on the outside of the house—without Grant's help this time because I didn't want a repeat performance of the previous year's debacle. He was still living with his girlfriend, and because of his behavior on Thanksgiving, we invited him over to open gifts on Christmas morning but told him he needed to find somewhere else to eat his holiday meal.

I felt conflicted about his not being there. Part of me was relieved—as were Taylor and Joan—that the remainder of our Christmas Day was markedly more pleasant without Grant upsetting any of us. But Joan, who always wanted everything to be peaceful and joyous, was also very sad that the whole family couldn't be together. It bothered me to see Joan and Taylor so distracted, wondering if Grant was celebrating somewhere with a Christmas tree and if he missed being with us, all of which made me anxious because I couldn't fix this.

It was a family tradition to take a photo of Grant and Taylor lying under the tree with all the gifts and discarded wrapping paper and bows. So Joan, trying keep with the annual ritual, had me lie down with Taylor, and as we were laughing she snapped our picture. Still, it wasn't the same without Grant.

23

TAYLOR HAD APPLIED TO the Fashion Institute of Design and Merchandising in Los Angeles and was ecstatic when she got accepted that summer. Watching her jump up and down, yelling, "I've been accepted! I've been accepted!" I figured that the emotions she showed after meeting the first of many goals she'd set for herself must have been similar to how I'd felt when I'd won my football scholarship to NIU.

Then it hit me. *Oh, God, now I've got to pay for this.*

Despite our financial problems, Joan and I still felt strongly that we wanted to pay for her college education, so Taylor and I sat down one afternoon while she was on Christmas break to come up with a game plan for getting her some financial aid. Her off-campus housing and tuition were going to cost $60,000 for the two-year program, and we certainly didn't have that kind of money anymore.

Seeing that neither of us had experience with seeking financial aid, I thought it would be a good lesson that both of us could learn together. Taylor pulled a chair up next to me in front of my oversized computer monitor, and we Googled *financial aid for college*. We clicked on the sites that listed programs offered by the federal government, which we both felt were a good place to start.

I sensed that Taylor wanted to take the lead in this project, and because I was feeling inadequate, I was all for it. She kept rattling off terms such as "FAFSA" and "ACG," which meant nothing to me.

"What's that?" I asked.

"It stands for Free Application for Federal Student Aid, and Academic Competitive Grants," she said.

I knew what student loans were, but I didn't realize that scholarships and other funds were available for students who weren't athletes, such as Pell grants for low-income students. As Taylor explained this to me, I loved letting this seemingly worldly high school student be my teacher.

We found the various applications online and plowed ahead despite my fear that I wouldn't know the necessary information. Once we saw the list of questions, Taylor tried to reassure me.

"Oh, this is easy," she said.

Most of them were in fact easy, but when it came down to financial queries such as "mother's and father's income," both of us were lost on where to get the answers. I knew Joan and I had made very little this year—only the salary that Joan had earned during her seven months at the hospice.

"I have tax returns; would it be on that?" I asked Taylor.

"I don't know," she said. "I hope so."

I pulled open my desk drawer, found the file of returns, and the figures were right there. All we had to do is transfer them onto the application. Under "projected income" for me in 2010, I put 0 because that's what I'd made in 2009. When it got kicked back, asking for 2010 projected income, I entered the same information.

Every time I looked over at Taylor, I could see the passion, the determination, and the drive on her face. Joan had told me and I had observed for myself that Taylor was creative and self-disciplined. She knew how to meet a deadline, and she worked well under pressure. I figured I must have had a lot of self-discipline to train and get good enough to play football, so I hoped she'd gotten at least some of those traits from me. I'd also noticed that she made up checklists, just like Joan said I used to do while running the aviation business, a tool she must have picked up while she and Grant were helping out, cleaning and restocking the aircraft.

I'm sure most fathers would have been able to complete the applications without their daughter's help, but I wouldn't have changed our time together that afternoon for anything.

At seventeen, Taylor seemed so mature. I tried to imagine Joan at this age, knowing that I didn't meet her until she was eighteen, after she'd already started college.

Was Joan just like this? Was she this mature and patient with her father?

Taylor and I discussed her "visual communications" major while we were filling out the applications.

"Tell me again what that is," I said. "What exactly do you want to do with visual communications?"

She explained that she would study how to strategically design displays, placing and dressing mannequins, for example, to create an environment in stores that enticed people to buy things. But she had bigger plans and higher aspirations than that.

"I want to be a stylist like Rachel Zoe on TV," she said, referring to the fashionista to celebrities who had her own reality TV show.

I didn't know who this was, so Taylor said she would show me an episode that she'd saved on the DVR, which would give me a firsthand look at her dream career.

I agreed, but I dragged out our time at the computer as long as I could. I'd wanted to look smart for Taylor and show her that I could teach her something. I'm not sure I succeeded, but I tried to convey the message that even if you don't know something, you can figure it out by yourself or by putting your head together with someone else.

Afterward she led me into the family room with some kettle corn and put on an episode of *The Rachel Zoe Project*.

"Let me show you what I want to do," she said, revealing a whole new world that seemed a little over the top, considering the flamboyant way the characters talked, dressed, and carried on. I could never picture Taylor like these people, but if this was what she wanted, I wasn't going to judge. Okay, maybe a little.

"Taylor, really?" I asked.

"I love it!" she said.

"They're running around like they're crazy. Can you do this?"

"I hope so!"

Taylor gestured in her usual dramatic manner as she described what was going on and how it fit with her goals. "Oh, you've got to watch this part!" she kept saying, just like her mother did when we watched TV or a movie together. I found it interesting that Taylor and I were both using the same medium of TV to learn about the world.

It was a bittersweet afternoon. With only ten months before she was going to move to Los Angeles, I was determined to enjoy every remaining minute with her that I could.

Once we learned that all her loan and grant applications had been approved, I felt proud. Even though I no longer had an income, I'd done my part by ensuring that her education was paid for.

.   .   .

Come January 2010 we finally sold the boat for $215,000, which was $35,000 less than the offer we'd turned down a year earlier because we'd thought it was too low. We took a pretty big loss over what we'd paid, but the economy was still terrible and we really needed the money to live on.

Finally feeling ready to get back to work, I switched over my virtual office so that my business calls went to my cell phone instead of going straight to voicemail. I also started doing some consulting work for a Brazilian company that wanted to buy a Learjet 60 midsize corporate jet, a gig I'd gotten through a new networking acquaintance of Joan's. I still wasn't capable of doing my old job, and Joan was concerned there was too much liability for me to even try, but I felt able to handle the more limited task of selling a plane, so I gave it my best effort.

I found a company in Switzerland that had a jet but was in default on its loan, which looked like the perfect match for my client. Problems arose, however, when the Brazilians wanted me to give them all kinds of information on the plane before we'd signed a contract. After seeking the advice of a couple of former colleagues, I realized that if I complied, the Brazilians wouldn't need me anymore. As a result we couldn't reach an agreement and our relationship ended in a stalemate. No sale, no commission.

I felt like a failure. The old Scott surely would have closed the deal. But after giving it some thought, I decided that overall it had been a good learning experience about what to do and not do. Bottom line was that I needed to study the business much more before trying that again. The problem was that I'd lost all my previous passion for the aviation business.

After spending a third of my life playing football, I'd lost my passion for the game too. I wasn't even a real fan these days. I was more interested in what the game used to mean to me and my teammates and what it still meant to society in general. I didn't see professional players acting like role models as they did in the old days. Too often they got into the news for disrespecting coaches and referees, doing drugs, and getting arrested. I still believed that playing in the NFL should be viewed as a privilege and a badge of honor and that players should show more reverence for the game.

But as careers, football and aviation were of no more interest to me. Today my professional passion lay squarely in sharing my story, trying to make a difference and achieving a new form of success.

. . .

We'd been through two Christmases now, and I still hadn't seen snow anywhere but on television. With both of us having little success finding work, Joan and I were going a bit stir-crazy, so we watched the weather forecast closely for the mountain town of Flagstaff, Arizona, which was a few hours' drive away.

I'd watched the winter Olympics on TV, which looked like fun, and blizzards and twenty-car pileups on the news, which didn't, but I was still very curious to know what snow felt, smelled, and tasted like.

"Is it safe to drive when it snows?" I asked.

"Well, yeah, it depends on how much it snows," she said. "Sometimes you need to have chains on your tires."

I didn't like the sound of that at all, but Joan assured me we wouldn't go unless the conditions were safe.

Some of my most memorable moments of discovery in the past year, when I'd experienced appreciation for the beauty in nature, had come when we left town. In Oceanside I'd seen the ocean and made love to Joan, all for the first time. In Hawaii we were surrounded by pineapple fields, coconut and palms trees everywhere you looked, and I learned there were a zillion varieties of tropical flowers. When we went snorkeling in the clear, warm turquoise water, the most gorgeous tropical fish swarmed around me: neon yellow, black with white spots, and long skinny noses. I only wished I'd spotted one of the sea turtles that Joan had described.

On a headache-free day when the road to Flagstaff was clear, we booked an overnight stay at a bed-and-breakfast there and set off on our adventure. I'd looked on MapQuest and found that the route was a relatively straight shot north on Interstate 17. What MapQuest didn't mention was that this was a twisty four-lane road of switchbacks as the elevation climbed from 1,000 feet above sea level to 7,000 feet at our destination. Nor did it mention that the scenery would be absolutely breathtaking.

I'd taken my first airplane ride when we'd flown to Dallas in the spring of 2009, turning to Joan when I felt a painful sensation, as if my ears were closing in on themselves. "What's going on with my ears?" I asked her. "They're starting to hurt."

Once Joan explained that all I needed to do was swallow to clear them, I settled down.

As we were driving up an incline toward Flagstaff, I now knew why I had that feeling in my ears again. "My ears are popping because we're getting higher," I said, pleased to have a chance to apply what I'd learned.

Surrounded by canyons on either side, I'd been watching the signs marking the increase in elevation and opened the window. We were doing seventy-five miles an hour as the frigid air rushed in. "This is great," I said. "We don't need air conditioning."

Sure, I'd seen different landscapes from across the country on TV, but I'd thought that Arizona, at least, was all the same—hot, flat, and dry desert. But outside my window, I'd been watching the sand, cactus, and brush slowly disappear and fill in with increasingly green canyons and rocky mountains in all different shades of gray, brown, black, and the rusty reds of Sedona. There were also patches of white stuff on the ground, which I assumed was snow.

When we were almost to Flagstaff, we pulled off at the scenic turnout to take a look. As I stepped out of the car, I immediately realized that, unlike watching TV or looking at a picture book, I could look way down into the valleys, which started with greenery at the top and ended in beige rocks and gravel below. I discovered in the process that I didn't like heights; they made me feel queasy and unsettled, which may have been partly caused by the loss of vision in my right eye. The wind was blowing, and I was worried that if it didn't blow me over, the ground was going to collapse under me and I might tumble all the way to the bottom.

The mountains in the distance seemed flat and faded while the closer ones looked more vibrant and colorful, and I could see the sharper details in them, the jagged rock formations that drew stark lines against the cloudless sky. "This is amazing," I said, marveling at the grandeur that surrounded me from every angle. "You can't see this in a picture. This is something you have to see in person."

Joan had already explained depth perception to me in relation to my vision loss, but I'd never been able to appreciate it from such a dramatic and magnificent vantage point. "You know there's a God when you look at something like this," she said. "Only a God could make this."

I felt small compared to the vastness of the scenery around me, overwhelmed by it all. After watching so many science programs on TV, I knew that rivers had carved out the Grand Canyon many centuries ago and that if I asked a scientist about the landscape before me, he or she would probably offer a very different origin than Joan's. I couldn't relate to what she was saying, but I didn't want to ruin the moment, so I just went along.

"Okay," I said.

As we crossed the Flagstaff town line, the ground was now completely blanketed with white and the forests of pines were so thick you could barely see between them. We pulled off the right shoulder of the road in an area that had been plowed, where it was clear on my side of the car but not on Joan's.

I walked around the front of the car and heard an icy crunch under my rubber-soled suede shoes as Joan took off toward the trees. I followed, sinking deeper into the snow up to my knees. In the clearing ahead, three families with small children were sliding down a hill on a cardboard slab, whooping and laughing with simple pleasure.

"What do we do now?" I asked.

"Just keep walking," Joan said, pointing at the footsteps in the snow in front of us. "They did it."

We walked another hundred yards or so to a place where the air was completely still, silent, and wonderfully peaceful. All I could hear was the breeze whistling through the trees and the occasional soft thud of a pinecone falling to the ground, sounds of serenity I'd never heard before. With all my senses on high alert, I watched a couple of birds soar overheard, bigger birds than I was used to seeing at home. They were also flying lower, so I wondered if they were looking for rabbits or squirrels.

"Are there bald eagles up here?"

"I'm not sure, but they might be hawks or falcons."

I craned my neck, gazing up at the treetops, which had to be hundreds of feet high, several times taller than the palm trees I was used to seeing. It was probably thirty degrees out, but the sun was shining, and the reflection off the snow was so bright I wished I hadn't left my sunglasses in the car. Although I knew I'd grown up in the snow, nothing felt familiar, even to the touch. Still, I wondered if perhaps my remaining emotional or muscle memory might explain why I felt more of a connection to the snow than I had to any of the sights in nature that I'd seen in Hawaii or Oceanside.

The snow felt cold and wet but softer than the ice cubes we had at home. As I picked up a handful, following Joan's lead to pack it into a ball, she threw hers and hit me in the chest.

"This is what you do, huh?" I said, feeling a rush as I threw one at her, hitting her in the back as she turned to avoid the blow. We both laughed, pushed each other down, and thrashed around like two little kids. I grinned as I imagined doing this many times growing up.

Besides reveling in feeling like an excitable child, I discovered some natural wonders that day that were more amazing than any others I'd witnessed since my accident. It was a most welcome change of scenery.

. . .

In early 2010 Grant called one evening, looking for Joan. When I explained that she was at a charity event and Taylor was working at the restaurant, he asked if I was home alone.

"Yes," I said.

"Motocross is on. You want me to come over and teach you about it?"

I had watched motocross several times, but I wanted him to explain it so I could see his passion and intelligence at work in something we had done together. But as inviting as this sounded, Grant didn't usually ask or offer to come over unless he had an agenda, so I couldn't help but wonder. Still, I hoped that this time his motive would be pure and we could have a quiet visit. I really wanted to feel the love for him and his love for me that we must have shared in years past.

"Yeah, sure," I said cautiously. "Come on over."

Once he arrived, we hunkered down in the family room. We turned to Fuel TV, and as he leaned toward me over the arm of the couch, Grant began to narrate what we were watching, starting with the necessary preparations before a race and relating it to what we used to do together.

"You spent most weekends with me driving the RV, grilling out, prepping and fixing the bike, and 'coaching,'" he said, giving me a playful jab about my habit of telling him what to do. He went on to explain how I'd learned everything we'd needed to know to take care of our own maintenance of the motocross bikes, from changing tires and blown clutches to all the brake pads he'd burned through. Joan had explained that I'd taught myself how to change out a top end on the engine in just thirty minutes rather than have to rely on someone else to do it.

"You always gave me crap about going through so many brake pads," he said.

"Did you *need* to use the brakes so much?" I joked.

I was really enjoying this. This was the first and only time we'd spent an evening alone, laughing together, since my accident.

This is how Grant must have been before my accident and his drug use. This is the loving, caring Grant, the one who wanted to be part of the family. The Grant Joan has been telling me about.

When the race started, he explained the riders' strategies and the importance of getting a good jump on the others. Very focused, he explained every facet of the race from the rules down to the black-and-white checked flag they waved when the winner crossed the finish line. As he threw around terms such as *throttle speed, weight distribution,* and *timing,* he sounded like an expert, and for the first time I felt proud of my son for doing something positive.

"How did you learn so much about racing?" I asked.

"We learned it together by asking and watching other riders, and some lessons you paid for—oh, and my natural athletic talents, of course," he said, grinning.

Hearing this—and the fact that we were able to banter with each other a bit—helped to offset some of the feelings of guilt and doubt I'd had about my parenting history. Joan had told me that the old Grant—a good, sensitive, and smart young man—was still somewhere in this confused and lost spirit, just like the old Scott was buried somewhere in me. The question was how to help Grant bring out the best in himself. I'd only seen glimpses—his mature conversations with Joan's parents and the genuine kindness he exhibited as he helped his grandfather out of a chair or took a bag of groceries from his grandmother's arms. The fact that he usually seemed to do this more for the benefit of those outside his immediate family made me wonder how we could motivate him to treat us the same way.

It was a wonderful evening, reliving some of the good times that I wanted so desperately to share with Grant now. It gave me new hope.

. . .

Unfortunately, Grant couldn't stay out of trouble for long. His girlfriend eventually kicked him out, so he came back to us, his tail between his legs. Once again, not knowing what else to do with him, we put him into another treatment program.

Two weeks later he stormed out of a counseling session saying he wanted to get high and kill himself. The staff threatened to call the police.

During a meeting with Joan and his counselor the next morning, Grant asked her to return his cell phone and wallet because he wanted to quit the program. Joan said no, wanting him to stay put.

Grant left the facility on foot, so Joan raced back to our house to safeguard the items. The staff reported his suicide threats to police, and an officer picked him up as he was walking toward his sober-living home. She gave him a ride there, where he called us, asking again for his things, including his unemployment debit card.

Well, I wasn't going to make this easy for him. "I'll use what's left on this card to repay some of the money you owe us," I said.

"No you won't," he said. "I have the police here, and they'll come and make you give it to me."

"Fine," I said. "Bring them here then."

Grant showed up at our front door ten minutes later with a police cruiser at the curb. Apparently the officer had agreed to help him collect his things because he didn't seem suicidal. With Joan at my side, I told Grant to go get the police, standing firm in my hard-ass routine, so he brought over a uniformed female officer.

"Grant is here to pick up his debit card, and it's in his name," she said. "Would you please give it to him?"

"No," I said. "He's going to go buy drugs with it. He's a heroin addict."

"Look," she said, "he's not high now, and we can't do anything but ask you to give him his card back. If you don't, he could file a police report stating that you're withholding his property."

This made no sense to me, but I felt I had no choice. "Just follow him, because he's going directly to his dealer," I said. "That way you can bust both of them."

The officer nodded politely. While Joan calmly told Grant that he should think about what he was doing to himself, I angrily went to my office to retrieve his things, which I reluctantly tossed onto the ground in front of him

"Here's your debit card," I said, frustrated. "Get the hell out."

"You need to get help," Joan said. "Call your counselor. We love you. Call us."

I wasn't so forgiving. I couldn't believe my own son had brought the police to my house. I'd been following Joan's cues for fifteen months, but after watching *Intervention* and *Celebrity Rehab* religiously for many months now, I could see that being so emotionally and financially supportive was no longer the way to go. So, as far as I was concerned, he had just severed all ties, at least with me.

24

THE PROBLEM with having no past was that I had to rely on others to tell me about it, but I also needed a photo of a place or event that they described to make it more real. What helped me even more was to actually stand where I'd stood when an important life event had occurred and walk myself through the motions of the original experience so I could re-create those moments. After getting my SPECT scan diagnosis, I'd buried my hopes for more flashes of childhood memories. But by seeing the places where they had occurred for myself, smelling the air, and visualizing what had taken place, I could imagine the rest. With Joan at my side, I felt that re-creating the most meaningful memories of my past was as close as I was going to come to recovering them.

So far, I'd been to Hawaii, Dallas, Oceanside, Palm Springs, and San Diego, and now it was time to retrace my roots in Chicago. Joan and I had been talking about taking a trip there, to see where our life together had started, but we wanted to wait until the weather was bearable. Once we'd made it through the winter months, we set off for the Windy City on Friday, March 19.

We landed at O'Hare that afternoon, and as we were heading into the city in our rental car, I immediately noticed how different the landscape was from Phoenix. Dismal and dirty, this was a world of gray. Not a single tree was in bloom as we drove west on I-80, parallel to the railroad tracks, where garbage was strewn about and black smoke spewed from factories.

Once we arrived at the Westin Hotel, on the Magnificent Mile, I was struck by how grand everything was. The skyscrapers were so tall I couldn't even see the tops, which disappeared into the white cloudy sky. But I loved the hustle and bustle and wanted to get out into it right away. As we made our way down Michigan Avenue, people scurried around us while we gawked at the upscale window displays in department stores.

I also noticed the contrast in weather, not just from Phoenix, but even from when we'd left the hotel an hour earlier. The long-sleeved shirt I'd worn was fine at sixty-four degrees, but now the temperature had dropped to thirty-nine and the wind had also picked up. Joan had told me about Chicago's strange weather, but now I could feel the goose bumps for myself. I could see why we'd moved to Phoenix.

We went back to the hotel to grab our leather jackets before flagging a cab and heading to meet my cousin Brad for dinner at a steakhouse. I enjoyed Brad's company and laughed at his stories, but since getting to know him again I hadn't been able to shake the need for caution I felt around him.

Over dinner we discussed our plan to visit my old neighborhood in Calumet City. Brad warned us to be careful because things had changed there since I left. By the time I was in college, my parents had moved to another part of town. "I wouldn't go down there unless you're packing a gun," Brad said. "You don't want to remember it like it is now."

Brad said he regretted missing our wedding, but my mother and his father had been at odds. He also explained that I was supposed to have been a groomsman at his second wedding, but I couldn't get away from work.

"I wish I would have been there for you too," I said.

After dinner Joan, Brad, and I visited a couple of different restaurant bars along Rush Street then put Brad on a train to Naperville and headed back to the hotel around 1:00 A.M.

As I lay in bed that night, I wondered what the next few days would bring. I knew I had another day of light fun before getting to what I knew would be the most challenging part of the trip: visiting Taryn's grave in Skyline Memorial Park.

. . .

When we woke up on Saturday, an inch or so of white fluffy snow had fallen. The wind was blowing, lowering the thirty-five-degree temperature by at least ten degrees, which helped me understand the true meaning of wind chill. When we began to get pelted by big drops of wet snow, I was prepared. Wearing the black scarf that Taylor had made especially for me, I zipped my jacket all the way up.

While Joan was having lunch with her friend Manny, an NIU cheerleader she'd coached in 1986, I met up at Ditka's Restaurant with Brendan Dolan and Darren Stahulak, who had played on the offensive line with me in college. Although I'd talked to Brendan on the phone several times since my accident, this was my first time meeting him in person. We'd been friends since college, Joan said, and I soon found out why.

As we racked up a sizable bar bill over the next six hours, we chased away more than a few patrons with stories of our college football days. It was good to hear a different perspective on my past, which Joan had been unable to show me, and experience some more of the male bonding I'd heard so much about.

Before we knew it, it was eight o'clock, and I'd had far more to drink in that one sitting than at any other time since my accident. Joan and Manny joined us for another hour of laughs before she and I headed over to Carson's Ribs to put some food in my vodka-filled stomach.

. . .

We managed to sleep in until 9:00 A.M. Sunday, when I woke up with an empty feeling. I knew it was going to be a difficult day for me and even more so for Joan. I was finally going to see the grounds where one of my children now lay, and it really bothered me that I had no memory of placing her there, let alone visiting her on birthdays or holidays. I'd never asked Joan about this, but I imagined that we'd visited her grave every couple of months during the five years before we moved to Phoenix in 1993.

We checked out of our hotel to move closer to the south, had breakfast at a pancake house, and headed to Tinley Park, where Joan and her two brothers had grown up in a modest house. I'd been expecting something bigger, but the squat ranch-style house was only about two thousand square feet, including the basement. Joan said the new owners had left the caramel-brown brick but had repainted the blue wood siding a sage green and also had replaced the garage with a carport. Otherwise, she said, the working-class neighborhood looked just like she remembered.

She showed me the schools she attended, the route she walked each morning, and all her friends' houses. Her eyes were bright with excitement as she recalled playing in the park and described a bit about each neighbor.

As far as I was concerned, my life hadn't begun until I'd met Joan, so my top priority was to see where I'd met, proposed to, and married her. So from there we drove to Zion Lutheran Church, where we'd had our wedding. As we pulled up to park, I recognized the beige brick building from the footage my mom had sent us. We entered through a side door and saw that people were settling into the pews. A service was about to start.

"They haven't changed a thing," Joan whispered as we made our way up to the front.

Joan guided me to a seat right behind where we'd said our vows. As I listened to the music and the pastor speaking, I pictured us standing there facing the front with the groomsmen to my right and the bridesmaids to her left, then closed my eyes. After watching the video, seeing all the photos, hearing Joan's description, and now sitting in the church, I was able to visualize the two of us looking into each other's eyes and promising, in front of our family and friends, to cherish one another until death do us part. For the first time in my recovery, I was able to connect with an emotion about a past event. With our wedding photos and video in mind, I felt just as I must have at a fresh twenty-one all over again. It was an incredible moment for both of us.

We left the church out the main doors, stopping to take a photo of ourselves where the guests had thrown rice on us as we'd emerged, holding hands, on our way to the limo. Joan and I, still holding hands, looked at each other and smiled with the mutual understanding of what I was experiencing—feelings of gratefulness that she was still standing right next to me twenty-five years later.

. . .

Next Joan suggested we find a White Castle to sample one of our favorite shared teenage treats: sliders, greasy minihamburgers, served with pickles, ketchup, and mustard. Apparently I used to eat eight or ten at a time. But now I was sickened after only one bite.

"I can't eat this," I said, thankful that we'd already arranged to get some pizza and buffalo wings with her niece Julie.

Finally it was time to go to the cemetery, which was surrounded by open fields. Joan and I both have plots there, and Taryn was buried at the foot of them, near a fifty-foot-tall oak tree, about ten feet from a private road. Its branches were bare, but I could tell that in summertime its leaves provided some nice shade for Taryn. Joan's grandparents were also buried in another section of this well-groomed, peaceful resting place.

It was windy and cold, in the upper forties, and the sun was trying to peek through the clouds as we parked the car. I took a deep breath, let out a long sigh, and opened the door.

"There's Taryn," Joan said, pointing.

We made our way to our child's grave, which was marked with a bronze plaque. A small bronze vase on a chain was turned upside down to keep debris from collecting in it. Wearing jeans and dress shoes, I knelt down on the hard, wet ground, and with my gloved hands I brushed the dirt and dried leaves from the plaque. I'd seen a photo of it in the Taryn scrapbook, but touching the frigid metal surface helped me connect the emotional loss of my daughter to this tangible place.

The plaque read, "Our daughter, So small, so sweet, so soon, Taryn Blake Bolzan, February 27, 1988," and knowing it hadn't changed in two decades helped me understand what I must have felt being there before. I turned the vase right side up and potted the small bouquet of green and white carnations the groundskeepers had left there as a memento on St. Patrick's Day.

"Does it feel like it just happened, or does it feel like twenty-one years?" I asked Joan.

"It feels like it just happened," she said softly, breaking into tears.

I hugged her and held her for twenty minutes, listening to the wind blowing and the birds chirping. Joan showed me where we'd once hung a bird feeder, made by Joan's dad, over Taryn's head, where it drew sparrows and cardinals. She also told me that the groundskeepers had removed the wreath blanket that had been placed over the grave during the winter and would replace it with a new one next winter.

Feeling the chill in our bones, we got into the car, turned on the heater, and sat for twenty more minutes, talking about the memories that my accident had taken from both of us. Because Joan had been under sedation for the C-section, she couldn't remember much of Taryn's christening, her baptism, or the funeral arrangements I'd made and told her about afterward. Because I'd been the sole keeper of those memories, they were now gone forever.

"They were just so precious," Joan said. "It was a once-in-a-lifetime thing, and they're gone."

Although I knew there was little or no possibility these memories would come back, I couldn't help but hope that one day I would remember seeing my daughter in the hospital with my own eyes, holding her lifeless body outside the recovery room, and sharing those painful memories of that tragic day with Joan.

. . .

The mood was heavy as we drove for about a half hour to Thorn-wood High School. Because it was Sunday, the gates to the main part of the school were locked, but to our surprise, the giant indoor track and field, where I'd wrestled, played football, and thrown the shot put and discus, was open. We walked through the halls of my old school and stumbled into an unlocked gym.

At first I didn't want to go inside. "I'm not going in there," I said. But Joan thought it would be fun. "Let's just go in," she said. When I wouldn't budge, she disappeared inside and came out a few minutes later.

"You've got to come in here," she said, beaming.

"No," I said, feeling annoyed. "Why are you doing this?"

"You have to see," she said. "Your name is hanging on the wall!"

Well, that got my interest, so I followed her into the basketball gym and broke into a big grin. There on the wall I saw a long narrow yellow banner with my name spelled out in black block letters along with the names of seven other alumni who had gone on to play professional football and baseball.

"*This* is why we go exploring," she said proudly.

It felt funny to get excited about seeing my name on this banner, but I couldn't help myself. We spent another half hour walking the halls and looking at all the historical sports photos and champion-ship trophies behind the glass cases. We never found the football memorabilia, which must have been in another building.

From there, we drove past my childhood apartment complex in Calumet City and headed to the area where I'd lived in high school and visited during college. From seeing pictures, I recog-nized the dark brick triplex, which we'd called a three-flat. When we got out of the car, I saw a guy parked in front of us get out of his vehicle with a gun handle sticking out of his back pocket, so we quickly looked around, took some snapshots, and jumped back into the car.

I felt no emotional connection to this place or to its run-down strip malls filled with pizza joints, fast-food fried chicken and fish places, and check-cashing storefronts. People didn't take care of their lawns, and everyone parked their junk cars on the streets, which gave it the feel of a poor, depressed neighborhood. I felt less connected to my childhood haunts than to Joan's in Tinley Park, perhaps because she didn't know enough about my years in this area to tell me any stories. Nevertheless, I didn't feel I was missing anything.

I was happy to leave for my folks' apartment in Orland Park, about twenty-five minutes west on I-80 and four miles from where Joan grew up. My sister Candi welcomed us into my parents' tiny but spotless two-bedroom unit, with its flowered couch and refrigerator plastered with family photos, where we talked for a while before going out for dinner.

Afterward, my mom's close friend Maggie and her husband, Paul, came over for a surprise visit and shared stories about me and their daughters, Lisa and Yvonne, who were close to my age and had been like two more sisters for me.

Maggie seemed to be the only one in the room who remembered that I'd lost my memory. My parents were still starting sentences with, "Remember when . . . ," but Maggie always said, "I know you don't remember, but let me tell you about . . ." I immediately liked her and thought she was a wonderful, funny, and warm person. Paul was much quieter, but I could tell they'd been very close to my parents for a long time.

My father shared some photos of his parents and two brothers that I hadn't seen before and also told some stories of growing up in a boys' school in Pennsylvania after his mother had died and his alcoholic father beat him and the younger of his two brothers—one time so hard with a metal pipe that he put my father in a body cast. After that episode they were made wards of the state. It was hard for me to hear these stories because I felt no child should have had to go through such pain, especially not my father.

Every time I got together with my dad I could see that I was a lot like him. He was a family man who wanted the best for his children and was truly in love with his wife and enjoyed the time he spent with her. I could tell that he was a gentle giant—someone happy with who he was and what he had done in life. It was always good to spend time with my parents and Candi to share new memories with them, but after such a long day, we were happy to hit the hay at the Westin in Lombard.

. . .

On Monday we took the one-hour drive to DeKalb to spend the day tooling around the NIU campus, search for the tree where I'd carved my marriage proposal to Joan, and meet up with Phil and Linda Herra. Phil had played on the offensive line next to me and was Grant's godfather, and other than Brendan and Jerry, he had been my closest male friend over the years.

Driving around the geographically expansive grounds, we found our respective dorms and saw the brick buildings where we'd attended classes, then parked the car at the lagoon where we used to picnic and feed the geese. The campus, which now has an enrollment of more than twenty-four thousand students, seemed much bigger than I'd imagined and was swarming with kids walking to and from class.

We strolled among the stand of white birch trees lining the grassy area around the lake until Joan stopped at one with a V-shaped trunk she thought was *the* one. She said she'd identified the tree years earlier, and although this one no longer had any visible marks, she figured either the carved bark had peeled off or it was too high above our heads now to know for sure.

Joan had told me the story early in my recovery, and as we stood there, the story came back to me: early that morning I'd told Joan I needed to run an errand. When I came back, Joan saw a screwdriver in my hand and wondered what I'd been doing.

"Let's go to the lagoon and feed the geese," I told her.

After gathering up some cereal and old bread, we headed over to the lake, where Joan saw me walking around, inspecting the tree trunks.

"What are you doing?" she asked.

"These trees are cool," I said. "Have you ever looked at them?"

Of course, she had no idea what I was doing, which was trying to guide her to the message I'd carved into the bark.

"Not really," she said.

When Joan still didn't find it on her own, I had to help her out. "How about this tree?" I said, pointing directly at the message: *Will you marry me?*

Joan smiled broadly and asked, "Are you going to get on one knee?"

"Of course," I replied, doing so. "Joan, will you marry me?"

"Yes," she said, tearing up with joy.

By standing in the very same spot where I had carved and said those important words, I was again able to visualize and re-create one of the most seminal moments of our life together. It almost felt like I was there doing it again.

From there, we met Phil and Linda at a greasy gyro place we used to frequent in the eighties. With all this unhealthy food I'd consumed, I was starting to see how I'd gotten into the bad eating habits that led to the gastric band surgery.

The four of us then went to Huskie Stadium, where Phil took me on a tour of the new facility that housed the football team and workout areas and also down to the field, where we'd played our games. Even though the scoreboard was new and they'd installed some different seats on the stadium's east side, I really got a sense of what it must have been like to play there. I smelled the mildewy scent of the artificial turf, looked up at the aging stands—red seats below and gray metal benches up in the cheap seats—and imagined a crowd of fans cheering us on, even in the icy winter months.

Phil also showed me the coaches' offices, the wall featuring the names of former Huskies like me who had gone on to play in the NFL, and the infamous forty-degree switchback ramps that curled around inside the stadium walls and also outside, where we had to do what I was told was my least favorite drill: running wind sprints during our off-season workouts. It really meant a lot to me that Phil had taken the time to bring me closer to him and my college football experiences.

After that Joan and I stocked up on NIU fan goodies at the bookstore, buying shirts, jackets, coffee mugs, and lapel pins, then we stopped for a taste of DeKalb's famous beer nuggets. Joan said we used to eat bagfuls of these deep-fried pizza dough chunks dipped in hot marinara sauce, which most girls blamed for their "freshman fifteen."

I slept well that night after such a memorable day. For the rest of my life I will cherish the new memories I made, walking the same footsteps that I had thirty years earlier, and I hope to return to watch some football games there in the future.

. . .

Tuesday was our final sightseeing day, and our first stop was in Glendale Heights, where Joan and I bought our townhouse and where we were living when we brought Grant home from the hospital. Like Grant, the trees that were babies back then had since grown up, although the townhome still looked like it did in the pictures I'd seen.

Next we visited the first house that we actually designed and built ourselves in 1990 in an upscale development at Aurora's Stonebridge Country Club, the site of many a high-profile golf tour. The house was surrounded with gorgeous oak trees and flowers, including tulips just like the ones we'd planted. Joan told me we'd held quite a few functions at the huge clubhouse nearby, including Taylor's christening party and the Easter Day celebration when Taylor decided she no longer needed her pacifier and gave it to the Easter Bunny.

. . .

Although I still didn't really know who I was, our trip to Chicago gave me a stronger sense of identity. The impressions I gathered by walking the streets, the playing fields, and the halls where I'd roamed all those years ago seemed to fit with what everyone had been telling me about my past. After I saw the neighborhoods where I grew up, where everyone worked hard and yet no one was rich, my consistently strong work ethic and the innate toughness I'd heard so much about now made more sense. Nothing had been given to me; I'd had to earn it all, and that was what had driven me to win and succeed, both on and off the playing field. I could also see how growing up in the Midwest, where family was so important, had instilled such a solid commitment to family in me.

Oddly enough, the pride I felt about my parents and where I came from seemed to belong more to the new Scott than the old one. Based on what Joan had told me, I'd never been a fan of Illinois, nor had I felt the need to live near my parents. I'd wanted to get away from there and make my own mark on life somewhere else, somewhere new. But going back there had made me realize how important it was to know where I came from; it had made me who I was—and could still become—even if I was living in a different state. I was a product of my parents, a working man and his loving wife, a mother of three. Why wouldn't I be proud?

25

FOR MONTHS before our Chicago trip, I'd been feeling hurt and confused by Jerry's failure to return my calls, so I sought a second opinion from my friend Mark. Mark had met Jerry on one of the half-dozen trips to Las Vegas that Mark and I took in one of my company's private planes, when the three of us played blackjack and drank all weekend.

"You know Jerry, right?" I asked.

"Yeah, I know him, not as well as you, but   "

"Well, tell me what you think of this," I said, describing my efforts to rebuild our friendship.

"That doesn't make sense," Mark said. "Maybe something's wrong—with the family, business is horrible, or maybe he just feels you don't know him anymore and he's uncomfortable."

"Why are you different?" I asked.

"To me, when a friend is needed, that's when you can best be a friend," he said.

I'd continued to leave dozens of phone messages for Jerry and also sent numerous emails saying I needed to talk to him. While we were planning our Chicago trip, I left messages saying we wanted to hook up with him. Still nothing.

From what I'd seen with Joan and Mark, not to mention on TV, friendships weren't supposed to work like this. Jerry and I had talked often for twenty years and told each other our secrets. So why, I wondered, this nagging silence?

Jerry's disappearance remained a regular topic of conversation for me and Mark. Neither of us could get over it. "You ever hear from Jerry?" Mark always asked, to which I replied, "Nope."

During our lunches and on a trip we took together to the boat, Mark and I often just sat and talked. He seemed to enjoy teaching me things and telling me stories about the old Scott. But Mark also seemed to enjoy the new Scott, the one who came to Mark's son's baseball games and exchanged stories about his personal life.

Given the disappointment with Jerry, it was nice that Mark still wanted to be my friend because that's what I wanted too—a friend who stuck with me, not out of a sense of obligation or guilt, but out of a desire to spend time with the new Scott.

Mark said our friendship had grown stronger and deeper than ever before, so, if anything, my accident had brought us closer. If he ever needed me, I would be there for him too.

. . .

Later in March, which was Brain Injury Awareness Month, I finally was set to meet Taylor Ward, the local teenager with amnesia, and her family, who had organized a "fun fair" to educate kids and parents about how to reduce sports injuries. There were about two hundred people at the carnival-like event, held on the athletic field at Walker Butte K–8 Elementary School, where her mother worked as a nurse.

I was really looking forward to meeting Taylor. Knowing what a difficult time I'd had finding answers and help for all my problems, I hoped to share some of what I'd learned. My motives weren't entirely selfless, however. Still struggling to find my own identity, I wondered if she could also help *me* learn some things about myself. We might be the only two people who had survived the same set of circumstances.

Mattie from the Brain Injury Association of Arizona had also asked me to represent the organization at the event, so I was wearing two hats. I took some deep breaths as I sat in the parking lot, wondering if Taylor's parents were going to accept me.

Are they going to want my help? Will they want me to paint a rosy picture for Taylor, or will they want me to be honest with her?

As I walked onto the school yard, I searched around for someone in charge and saw an athletic man in his forties who kept getting calls on his cell phone. I approached him, and sure enough, he was Taylor's father, Dave. After his wife, Kathy, hugged me, I began to relax, and we chatted, mostly about the abilities Taylor had lost, such as reading music, playing jazz, and riding horses.

"This is so important for Taylor for you to be here," Dave said, who was probably unaware that it was just as important for me to feel wanted by someone outside my family.

I said I was sorry for what had happened to Taylor, and although she'd lost her entire long-term memory, I noted that she'd really only lost twelve years because no one remembers much before age five anyway. Dave pointed out that she'd still lost touch with her entire life. "It's been a devastating loss for her and our family," he said.

Ironically, he said, her MRI and CT scans had both been normal, just like mine, so the doctors were wondering if her amnesia was psychologically based—the same thing they'd tried to tell me.

Being four months ahead of Taylor in my recovery, I thought that meeting her would be a good measuring stick for my progress. When she joined us, the first thing I did was to look into her eyes, the windows to her soul, to see if she looked the way I felt inside. Her blue eyes were framed by red hair and fair skin, and as I searched them I felt like I was looking into a mirror. She had the same blank expression that Joan said I had right after my accident. I sensed that she was putting up a front to hide her despair and hopelessness, and I hoped I could help her come out on the other side of that, as I was beginning to do myself.

Shaking her hand, I tried to convey my compassion and understanding. I noticed she stuck her hand way out to keep a good distance between us, but remembering how sensitive I'd initially felt about being touched, I empathized completely. At first she seemed distant emotionally as well, but I could tell that she was a sweet girl.

"What are you feeling?" I asked, repeating Mattie's helpful question to me. "Tell me what it's like for you."

"Lost," she said.

"You know, I couldn't pick a better word," I said. "That's how I feel all the time."

We didn't get a chance to talk much before it was time to do the raffle, with which Taylor asked me to help because she was feeling shy. When we got up on the bleachers to announce the winning ticket numbers, she turned to me and said, "Scott, you talk."

But knowing she needed to build her own confidence, I encouraged her to do the job herself. "No," I said, "it's your event."

We worked out a compromise: I picked a winning ticket from a plastic bucket, and she called out the number.

A little while later Taylor and I sat and compared stories, and over the next forty-five minutes she gradually opened up. When I asked if she still had any friends from before her accident, she said that all but her best friend had deserted her. "Once they saw me after the accident, all of them looked at me like I was from a different country, pretty much. How 'bout you?"

I told her I didn't feel a real connection to most of my old friends. The effort required to rebuild those relationships seemed too overwhelming, so the idea of losing the friends bugged me more than the friendships themselves.

"How are you not angry?" she asked.

"I don't know," I said. "You can't worry about other people. The people you need to focus on are your family and the people who have been there for you since this accident happened."

"I still don't view them as family," she said.

"Give yourself time; that may change," I said, suggesting that she trust her feelings and gut instincts, because that's all damaged people like us have to go on. "Do you have any doubts that these are your parents or brother?"

"No, I just don't know how I felt about them before."

"That's probably the one advantage I have," I said. "I have forty-six years of stuff inside me, you have seventeen. Let yourself develop into the relationships."

Instinctually, I believed that all my years of experience and thoughts were still in my brain somewhere, I just couldn't retrieve them, because Joan said I still reacted the same way in many situations, even using the same words. If we had an argument, for example, I shut down and got mad a day or two later, same as before. I figured it was similar for Taylor, only she was still a child inside.

It was rewarding when Taylor wanted to take a photo together, which she later posted on her Facebook page.

"This is one of my new best friends," she told the one friend who had stuck by her.

I let her know that she wasn't alone and that she could contact me anytime. "If you ever need help, just message me," I said.

Although I knew we both had a long way to go, I felt it had helped both of us to connect. From then on, I always tried to answer her occasional Facebook message within a couple of hours.

. . .

Our May 26 wedding anniversary was coming around again, so I posed the question to Joan once more about renewing our vows, and this time I suggested we go to Paris. I figured we could use the Starwood hotel points we'd accrued at our time-share in Hawaii to offset our lodging costs. I'd heard many times that Paris was the most romantic place on earth, and neither of us had been there. Joan not only agreed to the idea, she was ecstatic about it. But as we started planning the trip, we realized that the airline tickets we'd bought at an NFL alumni fund-raiser had restrictions that meant we'd have to go for an entire month or for less than a week, neither of which worked. Much to our disappointment, we had to call it off.

My next idea was to get remarried in San Francisco. Joan wasn't excited about going to the Bay Area, however, because we'd been there together a number of times before my accident, and she was hoping to go somewhere new for both of us, but she never revealed this to me.

Joan had stopped the Cymbalta in March, blaming it for her weight gain, and thinking that things were now less stressful. But she was wrong. As the date for our trip to San Francisco approached, I didn't notice how overwhelmed and upset she was getting or that she felt the trip was contrived, overplanned, and too expensive. With Grant's problems and all of this on her plate, Joan was heading into a downward spiral, which I discovered only by accident.

I went out the morning of June 21 to run some errands, but when I was a block away I realized that I'd forgotten something and had to go back for it. As I was heading toward my office, I heard Joan crying in Taylor's bedroom with the door open. Coming closer, I heard Joan complain that it was unpleasant to be around me because I was getting more upset about the little things, just like the old Scott.

"I love him, but I don't know if I'm *in* love with him right now," she said.

Oh, my God. The only person that I trust and love may not love me.

"We've *been* to San Francisco," Joan went on, "and I don't want to be a damn tour guide. He wants to hurry this, and I don't."

Holy hell. What did I do?

I realized I shouldn't be hearing this, and I didn't want to hear any more. Part of me wanted to say something, and part of me wanted to flee without a word. I went with the latter, but not before audibly shutting the family room door to let them know I'd overheard their conversation.

Angry and deeply hurt, I left the house to cool off and drove toward Tempe, with no particular destination in mind. Joan called four times within the hour, and when I wouldn't pick up, she finally left a message.

"Let's not play these games. Call me back. Let's meet and talk."

But I didn't want to talk about this. I couldn't understand how I'd left her in relatively good spirits that morning and it had escalated to this. She was so persistent, however, that I grudgingly agreed.

Over a confrontational lunch in Tempe, she expressed remorse at what she'd said, admitted she'd said things in anger, and tried to explain the reasons behind her outburst.

"I'm sorry," she said. "A lot of it is I'm not happy with myself. I need some help."

She tried to convince me that her words had just come out wrong, but I believed that people meant what they said. The tense debate didn't get us anywhere, so Joan tried again after we got home in our separate cars.

"I *do* love you," she said. "It's just, with all this stress, it's hard to feel that romantic love."

"I don't see the difference," I said.

Still confused and feeling the sting of her words, I turned inward. I was still angry, and I didn't want to say anything more because I was trying to process the hurtful things she'd said about me and her feelings, or lack thereof, for me.

How can you love someone and not be "in love" with them? Will this be a regular thing with her, falling in and out of love with me? And how can I trust that there aren't other things she's been keeping from me?

She spent the next two days in bed before both of us went to see Dr. Lanier. I waited in the lobby while she talked with Joan, after which I went in myself. Dr. Lanier recommended that Joan switch to Wellbutrin and start seeing a therapist, and within a week things were already better between us.

But I was still having trust issues. The things I'd overheard Joan say had been playing over and over in my head like a cruel tape recording.

We were watching TV one night a few days later when Joan got up from the couch and settled on my lap on my chair, and after a week of keeping her at arm's length emotionally, I let her cuddle with me. "I think I'm falling back in love with you," she said.

"Well, that's good," I said, not knowing what else to say.

"Do you still love me?"

"I was never out of love with you."

Slowly, I felt the gaping wound in my heart begin to heal. I figured the meds were what she needed, and this was something I could understand. Nonetheless, she'd have to do a lot more than this to get my trust back.

Throughout this entire episode, I had learned another lesson: you can't help anyone else until you help yourself. When an aircraft is in trouble, you have to put on your own oxygen mask before you put one on the person next to you. Neither Joan nor I could help each other as long as one of us was unhealthy and lacked the proper coping tools.

Our wedding anniversary passed without the ceremony that we'd hoped would commemorate it, but I expressed the same sentiments in a letter that I still hoped to say before an officiant sometime soon: "Even though I lost my memory of my past, there is one thing I do know, and that is my love for you can never be taken away by my loss of memory because it is in my heart forever. My heart is where my love for you will always be. . . . There is no one I love more than you."

I was still throwing out ideas about where and when to renew our vows. Our finances were no less of a mess, so we couldn't afford a fancy to-do. My latest idea was to return to the notion of a simple backyard ceremony with just the family.

After my accident, I'd forgotten what it meant to have a wife. Since then, I'd relearned not only that concept but also how to be a husband from the same partner who had taught me by example. To me, a good husband was the male equivalent of Joan—a loving, devoted, faithful, honest, and supportive gentleman. Someone who stood by his wife through the good and the bad so neither of us would have to be alone and who was there to share dreams, emotions, and thoughts without judgment. This template seemed to have worked for our first quarter century together; why not for the next fifty years?

26

WITH OUR RELATIONSHIP back on track, we headed to Santa Monica, California, for a few romantic days of relaxing and writing by the pool. The trip really did the trick; it even made us wonder if we should move there.

We were fifty miles out of Phoenix on the trip home when Grant called on my cell. I had the Bluetooth speakerphone switched on, and I knew something was wrong when Grant asked Joan to take it off speaker. Hearing her say, "uh-huh, uh-huh, uh-huh," and seeing the look of concern on her face, I knew what had happened.

"When did he relapse?" I asked after she hung up, telling her not to beat around the bush. If I asked, that meant I could take the bad news.

She explained that Grant had, in fact, relapsed several times recently and had been kicked out of yet another rehab program. His call to us today had come after his third relapse, following his moving in with a new set of sober friends from AA.

"All our relaxation and fun just ended with that phone call," I said.

By this point I'd decided to hold my ground about cutting ties with him, so I told Joan I didn't want to see or talk to him until he'd been sober for sixty days, and I didn't want her giving him any money, food, or help of any kind. If he couldn't make it to sixty days, then I would keep doubling the time until he could.

Enough was enough. "You can talk to him, but I don't want to," I said.

A couple of days later we met with Grant's counselor from his last program, and I told him my plan. He said it sounded fine and encouraged us to set and discuss those boundaries with Grant, which I did. Taylor didn't want to be contacted either, and Joan said she would accept a call from him on Sundays so she would know he was still alive.

I was dreading the approaching date of Taylor's departure for Los Angeles. Our bills were still mounting by the day. I could feel the pressure building in me, but even as I tried to keep moving forward, I was finding it much harder than usual to concentrate. Some internal pot seemed about to boil over. But instead of attending to it, I chose to press forward.

· · ·

I was trying to flush the negativity from my mind when an exciting opportunity presented itself. A businessman I'd met during Grant's motocross days approached me about buying the jet company—my contacts, trademark, software, website, letterhead, and brochures—with a couple of partners. We arranged to meet for a couple of days in Newport Beach, California. I insisted to Joan that I go alone, to prove to myself that I could do this.

I drove out to Newport, the meetings went well, and I was cautiously optimistic about the deal. We left it that they were going to seek some additional funding and get back to me.

On Wednesday, July 21, around noon, I was getting ready to return to Arizona when a crushing pain shot through my right temple and into my bad eye, causing my right visual field to blur and shift to the right. This was a new pain, stronger and sharper than my usual pain in the back of my head where my skull had split open.

"I've got a real bad headache," I told Joan on the phone. "I don't think I can drive right now."

Joan suggested that I go back to the room and stay another night, but I'd already checked out, so I told her I was going to rest in the lobby and see if the pain subsided. I took some medication, and the pain did improve within an hour, but I waited another hour just to be sure. I didn't want it to hit me again—or turn into something worse, leaving me stranded in the middle of nowhere. By 2:30 my pain was back to a bearable five out of ten, so I told Joan I was heading home.

. . .

For some weeks now I'd been feeling overwhelmed with emotion—fear, emptiness, hopelessness, and now anger—to the point where I could no longer control it, even with the Cymbalta. These feelings, on top of the pain of old and new headaches, were scaring me. I'd been trying to rationally think my way out of my fear, but nothing was working. I simply couldn't cope with everything the world was throwing at me. I felt myself slipping into that dark vortex of ruminations again.

What if Grant never gets better? I might never have an emotional attachment to my son. What am I going to do for income? How are we going to afford the mortgage and the cars? How will we afford Taylor's college education and ancillary costs if something goes wrong with the grants and the student loans? I feel like a failure—my wife has to produce an income because I'm not capable of producing anything. I don't know what to do.

I was out driving around, trying to sort through my feelings and having a very difficult morning as I was working on the book chapter about my felony. It was very agitating and upsetting to me to relive all that. At one point I felt so overwhelmed I had to pull my car over to the shoulder of Interstate 101, where I sat, crying.

I don't want to live like this anymore. I wish I'd died in my accident.

I got out of the car and stood next to it, my butt pressed against the door, about two feet from the line on the pavement that separated me from the cars and trucks whooshing by at sixty-five miles per hour. As I had done countless times since my accident, I visualized all the ways I could get myself killed in traffic by making it look like I'd had *another* accident. Bending over to tie my shoe and tumbling into the lane in front of a truck (never a car because I didn't want to critically injure myself and survive). Stepping up onto the rim of the bottom edge of the open car door, reaching for something on the roof, and pretending to fall backward. Or walking out into the closest lane, pretending to stumble and lose my balance, and ending it all.

But as I stood there that day, my thoughts eventually went to the same place they always went.

What about Joan, Taylor, Grant, and my parents, family and friends—how would they go on if I did this? What would it do to them? I could never put them through this. After all, I know what it feels like when Grant threatens to commit suicide. My stomach turns, and I feel a burning sensation in my heart. It would kill Joan if either one of us actually died.

So instead I got back in the car. Deep down, I felt there had to be a reason or a purpose for my fall, and although I wasn't sure what it was, I told myself that committing suicide before I could find out was not a good option. I willed myself to reach deep inside, to fight my way through these emotional lows, and to prevent my body from giving up the battle with my mind. But I honestly didn't know how many more times I could pull through one of these episodes, especially when I was reeling from this new crushing pain.

When I got home I took some medication and sat in my favorite chair. It felt as though someone had stabbed me in the temple and sliced into my eye.

Joan came over to ask how I was. "Why are you so upset?"

"I have a headache," I said dismissively, closing my eyes again. After about an hour the throbbing new pain diminished and I was back to my usual headache.

. . .

Joan and I escaped the Phoenix heat for my birthday weekend amid the cool pines of Prescott, a mountain town a couple of hours away on Interstate 17, where we stayed at the new summer home of our friends Dr. Rich and Kathy Silver and played a relaxing round of golf. The weekend was pleasant, but I felt as if I was just going through the motions, disconnected from everyone, and not really living in the here and now. I figured my strange feelings were caused by staying at someone else's home for the first time, so I kept them to myself.

On Monday, I met with my attorney in Scottsdale, and had lunch alone because Mark was tied up. I was driving along Shea Boulevard toward the 101 freeway heading home when I came to a stoplight with six cars ahead of me in traffic. To the right was the open desert, and it was hot out. I felt the pain in my temple come on again, only this time it erupted much faster and more intensely, as if a harpoon had pierced the side of my head. The next thing I knew, my car was stopped on the desert sand. The gear was in park, the engine was still running, and the radio was playing. But I had no memory of pulling off the road.

How did I get here? Why don't I remember what's going on? This isn't good.

Figuring that I had blacked out, I called Joan and told her what happened.

"Where are you?" she asked.

I told her I was near the hospital in Scottsdale and the office of Dr. Arlen, the neurologist who had ordered the SPECT scan, and was going to go see her. Joan said Dr. Arlen would probably just send me to the hospital for a CT scan, which is essentially what her receptionist told me to do. Joan asked if I was okay to drive, and I told her I could make it to the ER, which was just a few blocks away.

"How's your head?" she asked.

"It's killing me," I said as I approached the hospital where Joan said she would meet me in the ER's triage area. Inside the ER, a nurse named Deb, who had worked with Joan, recognized me and took care of me until Joan arrived. With Deb there, I didn't have to persuade everyone all over again that I wasn't crazy, that I really did have a brain injury that caused amnesia. Joan always said God puts people in the right place at the right time, and this was definitely one of those times.

You might think I'd feel a frightening déjà vu at being back in the ER with an excruciating headache, but I didn't. This time I knew exactly who and where I was. I didn't feel confused, and I was able to explain to Joan and the nurses what had happened in the car. I also didn't feel paranoid or all alone, like I had the first time. I'd come a long way in the past eighteen months, but I was still scared that my condition had taken a turn for the worse. I was relieved to see my wife walk into the ER and to receive her hug and kiss.

"At least I know who you are this time," I laughed.

When I became intensely nauseated all of a sudden, Joan asked a passing technician for a basin. She turned away to get a cool washcloth to put on my head when I started groaning, grabbing the right side of my head, and grimacing with agonizing pain. My legs shot out and grew rigid, with my toes pointed down, my arms flexed, and my elbows bent and pressed tight against my chest. My body began to shake all over and rock from side to side. I was having a seizure, which was like a foot or leg cramp you get in the night that leaves you really sore and limping for days afterward, only in this case the cramp had taken over my entire body.

Joan leaned over and pressed her arms down on my shoulders, trying to hold me flat and prevent me from hurting myself. But she was too small to restrain me. I flew off the bed toward her, landing face-down on the floor and smacking my jaw on the cold linoleum. They'd put me in a hospital gown by then, and although I would usually be happy to be wearing underwear so my naked butt wasn't exposed, at that moment modesty was the last thing on my mind.

"I need help!" Joan yelled, which brought five ER staffers running, including Dr. Robert Londeree, a longtime ER doc whom Joan had worked with when she was a nurse.

A male tech squatted at my feet with Joan and took my arm to try to help me up, but I felt so disoriented I thought he was attacking me. I twisted to sit up and drew my fist back to go after him.

"Whoa, relax," the tech said.

Joan quickly put her hand on my left shoulder to try to soothe me. "Honey, honey, it's okay," she said. "You're in the ER and fell off the cart. It's okay. They're here to help you."

Dazed, flustered, and flushed from the pain, I let them help me get up—slowly—and back onto the bed, this time with the side rails up.

Dr. Londeree greeted Joan with an awkward smile, given the circumstances, and after assessing me, he ordered a CT scan and some blood work.

"Can you give him something for the pain?" Joan asked.

The nurse put some morphine into the IV, and as the drug began to take effect, she faded away. Joan had to fill me in later because I promptly forgot most everything that happened for the next week, either because of my brain misfires, the heavy medication I was on, or both.

The doctors, who dosed me with painkillers and antiseizure medications that provided no relief from the symptoms, were unable to come up with a diagnosis. In between a series of tests that all came back normal, Joan said, I had bouts of confusion and disorientation, but I was also able to hold coherent and lucid conversations with her and the staff—at times cracking jokes left and right.

When, three days later, Barrow Neurological Institute wouldn't accept me as a transfer by ambulance, Joan had to take me there in her car, somehow managing to keep me from banging my head against the window when I seized up along the way.

I spent several days at Barrow, where on my worst day I had a whopping fifteen seizures, which terrified both of us and wore me out. I lay frustrated and scared that my life was over.

"What is happening to me?" I kept asking Joan. "Why doesn't anyone know what's going on with me? Is this it? Am I going to die?"

"We're trying to figure it out, hon," she said. "Hang in there."

After each convulsion subsided, I felt disoriented and weak. The pain was just too much to bear. "I can't live like this," I told her. "It hurts so much."

By Saturday, the least bit of activity triggered a seizure, whether it was eating, going to the bathroom, or brushing my teeth. Nothing seemed to relieve my pain, including a new drug cocktail. I was so gorked out, Joan said she had to feed me and remind me to swallow because I kept nodding off with food or drink in my mouth.

On Monday I was transferred to the coveted epilepsy monitoring unit, where they covered my head with sticky white electrodes and took me off the antiseizure meds. The plan was to keep me hooked me up to a video camera and EEG machine to monitor my brain activity and determine where in the brain the seizures might be originating.

At times I felt like I was losing control over everything and that Joan and everyone else were hiding things from me, another side effect of the drugs. And by Wednesday I was not only fuzzy and confused but also paranoid, sure the doctors and nurses were all talking about me.

"I feel like everyone is out to get me and I don't know where I am. I feel like I'm Jason Bourne," I said, referring to the movie character CIA assassin who wakes up one day and doesn't remember who he is but is rightly convinced that people are trying to kill him.

"Babe, I'm here and you're safe in the hospital," Joan said. "I won't let anyone do anything. I'm right here."

"I'm glad I got you."

Finally the moment of truth arrived, and without the antiseizure meds, I was feeling more alert and could remember things again.

After reviewing the EEG data and videotape with a team of experts, Dr. Jason Caplan, the chief of psychiatry, told us that my seizures were *not* epileptic, meaning that my brain showed no correlated electrical activity, but were of psychological origin.

"First," he said, "Mr. Bolzan, you're not crazy. This is not in your control, nor does anyone think you're faking it."

My attacks and pain, he went on, were caused by my brain converting severe stressors into a sudden onset of involuntary symptoms. In other words, after all my efforts to hold things together, my body had been forced to give up the fight with my mind. Essentially, I had lost the fight with myself.

In addition to losing my memory, he said, I'd had to deal with our financial issues, Grant's drug addiction, and new career challenges, so it wasn't surprising that I would feel overwhelmed. But instead of dealing with those stressors, I'd been pushing them down, ostensibly to keep moving forward. Joan and I had always told Grant that he was his own worst enemy, and sadly, I had to admit we had something major in common after all.

"It had to come out somehow," Caplan said.

My treatment, he said, would be to find a good therapist and learn some better coping skills. He gave us some names and said we could go home.

. . .

For the next two days I wasn't allowed to drive or be alone in the house unless I was in bed or in my big chair. It was frustrating to feel I'd regressed twenty months to being dependent on everyone again. Still very tired, I slept a lot, but as the frequency and intensity of the convulsions fell off, we both hoped this frightening episode was coming to a close.

When I woke up Saturday morning, I felt like something miraculous had happened. For the first time since before my $46,000 hospital stay (all but $2,000 of which would thankfully be covered by insurance), I had no headache. I can't even explain how relieved I felt when that gripping pain stopped. I bounced around the house with more energy than I'd had in weeks, so much so that I insisted on cutting the grass.

On Sunday I felt even better, well enough to accompany Joan to the grocery store.

"I can't believe that I don't have a headache at all," I said, amazed.

"See what a drug-induced coma for eight days will do?" Joan joked.

"I know, this is so weird," I said, my optimism in life renewed. "I feel great. Whatever it was, maybe it's gone and I can move forward yet again."

At long last—we hoped, anyway—the chains of pain had been broken.

Even though I'd had an extraordinary do-over in life, this new opportunity clearly came with its own unique challenges, not to mention the reality that I was going to have to deal with its physical and emotional ramifications—along with all the other stressors of everyday life—for some time to come.

That was the rhythm of my new life: two steps forward and one step back.

. . .

I ended up going back to see Dr. Barry, the same therapist I'd visited a couple of times but stopped seeing when I'd felt I had everything under control. Luckily, he wasn't an "I told you so" kind of doctor.

We agreed that I should see him weekly for at least three months, but he cautioned that it could take years for me to internalize what I needed to learn—how to relax, to identify stressors, to listen to my body and mind, and to make necessary adjustments before I reached another crisis point. But what I most needed was to increase my self-motivation and self-worth and try to stop feeling that I was failing myself or my family.

He predicted that my biggest challenge would be to adjust to who I was now and accept that I might never know who I'd been before, to face the reality that I had experienced a death—in myself. I should grieve the old Scott, which might take years, rather than try to bring him back. And he advised that instead of obsessing about what I didn't know or understand, I should focus on all I'd achieved since the fall *without* any knowledge.

No matter how strong or in control I thought I was, clearly my mind and body both needed to get stronger, just like when I played football. And the only way that was going to happen was by "working out" with Dr. Barry.

I just had to keep telling myself that I'd managed to retain some sort of innate strength and resourcefulness that would allow me to keep pushing through the roadblocks my new life kept throwing my way.

27

ONE FRIDAY AFTERNOON in late September, Joan and I drove to Tucson for a night away. We were within five miles of our hotel, leaving a restaurant after lunch, when we both checked messages on our cell phones. Seeing I had a text message from Grant that said, "I love you all very much. Goodbye," I took this as a nice sentiment—until I heard Joan exclaim, "Oh, my God, what does this mean?" We didn't know this yet, but Grant had sent this message to all the contacts in his phone, including Taylor.

Joan called and reached a groggy, incoherent Grant, who said, "I can't take this anymore. The next shot is going to kill me," and hung up.

In a strange technological coincidence, Joan screamed and started crying, not realizing that she had picked up a call from Taylor immediately after hanging up with Grant, and promptly hung up on our daughter. Taylor, taking Joan's reaction to mean that Grant was already dead, was in hysterics when she called me moments later.

This all happened in seconds, but by this point it was clear there was more to his text message than I'd initially thought, so I turned the car around and sped back home toward Gilbert as Taylor, Joan, and I tried to figure out where he was so the police could try to stop him.

To make a long story short, Grant had bought enough heroin to overdose, but thankfully he was too high after shooting up the first time to prepare to inject himself with the rest of his stash, which he then lost. We got the police to pick him up, to deem him a danger to himself or others, and to have him transported by ambulance to a mental health and detox facility, which held him for seventy-two hours.

On Monday Joan crashed my therapy appointment with Dr. Barry, where we spent ninety minutes discussing what to do and how to handle Grant's next announcement by phone, which was that he was moving out of the apartment he shared with three sober friends from AA. He'd decided to live on the streets and do drugs until he died.

"I don't want to hurt anyone anymore," he told Joan.

"Grant, please don't," Joan said. "We have lost so much—Taryn, Dad's memory, and we can't bear to lose you too, our only son. I love you so much."

But he was not in the state of mind to hear her.

By this point my anger, which had largely stemmed from Grant's disrespectful behavior toward me and his mother, had been replaced with compassion and a better understanding of my son's condition. Once Grant's suicide threats had become real and I realized they were a call for help, I dropped the sixty-day rule. Dr. Barry had helped me, and now Joan, to accept that my son had a serious problem with depression and that he'd been self-medicating with the heroin. But until he got the mental-health help he needed and stuck with the antidepressants, he wasn't going to get any better. If he wouldn't take the help we'd been offering or seek help for himself, he was doomed.

"It will never be your fault," Dr. Barry said.

As much as it hurt and saddened me to say this, I didn't think Grant would be with us for Christmas.

For months I'd been tormenting myself with my inability to help my son.

How can I expect to help others if I can't help Grant?

Dr. Barry said I should focus my efforts on trying to help those people who actually wanted and were willing to accept my help, but that didn't mean I didn't spend every day hoping that Grant would call to say he was tired of living like this, that he was ready now to do anything, anything at all, to get better.

. . .

Several days after the text message, Taylor and I were in her bedroom talking while she packed for the move to Los Angeles, and she told me she'd decided to cut ties with her brother. She couldn't take the pain anymore either. "What if I get another text message while I'm in LA and there's nothing I can do?" she said tearfully.

I tried to comfort her, to let her be the child and me the parent for once. She had clearly shouldered enough of this burden, and I finally felt ready to assume the parenting duties. My eighteen-year-old daughter had carried out the responsibility to the best of her ability, teaching me what she could along the way. "Let Mom and me take this over," I told her. "This is your time to live. I apologize for you having to see me in this condition, and I certainly apologize that you have to go through this with Grant."

"Please don't apologize for you or for Grant, Dad," she said. "Your condition is from an accident. Grant chose his life."

"Well, I apologize anyway because you shouldn't have had to go through all this. That's why it's important for you to go out to LA and not worry about this."

I offered her the option to block Grant's phone number from calling or texting her, and she said she would think about it. Although she hadn't talked to him for months before the text message, she ultimately decided not to make a more permanent break because she didn't want Grant to feel she'd abandoned him.

After our conversation I realized that I had passed a major milestone. Since my accident, I'd relearned what it meant to be a good father: a role model, a teacher, and a purveyor of right and wrong. But this was really the first time that I had been able to take charge and *be* a father, speaking from the heart with confidence and not needing to parrot what I'd heard Joan or some father figure on TV say. My words seemed to comfort Taylor because she folded into my arms like a little girl, which let me know that I had effectively resumed my job as the protector I'd once been.

I felt empowered, uplifted, and valued, thrilled to be able to do this for Taylor for the first time since my accident.

. . .

Taylor had graduated from high school in May, and I'd been pleased to be surrounded by Grant and both sets of grandparents as we watched this rite of passage for the daughter to whom I had grown so close.

When I saw her walk onto the stage in her royal blue cap and gown to receive her diploma, her image reflected on the jumbo television screen, I stood up with Joan and Grant to cheer and yell out her name, tearing up at the thought that this brought her one step closer to leaving home.

"I wish I would have remembered your entire high school education, but from what I've seen the past year, I can't tell you how proud I am of you and how hard you've worked," I told her afterward.

Taylor, who was beaming at me, responded with a hug. "Thank you," she said.

Now, four and a half months later, the painful process of letting her go was heading into its final stretch.

. . .

On Wednesday, September 29, I woke up at 5:00 A.M. with a knot in my stomach, knowing that this was the last morning I would see my daughter roaming around our house. There would be no more of her unique naïveté and what we called Taylorisms, the words she made up or messed up, such as saying "disposable" thumbs instead of "opposable" or that it looked "musty" outside rather than "muggy." Selfishly, I would have loved to keep her home and spend more time with her, but I knew she had to pursue her passion and become the person she wanted to be, just as Joan and I had done at her age.

She'd really been there for me since the accident, and I'm sure the past two years had been hard on her. She'd told me about so many memories of the two of us spending time together, like the father-daughter dance we'd gone to or when I'd helped her race her four-wheeled vehicle, known as a quad, during Grant's moto-cross days.

I would forever miss these special moments and could only hope that I'd encouraged her and shared enough of my own life lessons before the accident so that she would have some valuable advice to carry into her adult life. If Taylor could forgive me for not remembering her first seventeen years, I could promise to help create even better memories for both of us: her first day of college, walking her down the aisle, holding the first child I imagined she would bring into this world, and who knows what else.

As we finished packing the last of her things for the road trip, she and Joan hopped into Taylor's car, and I climbed into Joan's. Anthony, who was starting college an hour or so south of Los Angeles, in Orange County, was going to follow up the next day in a U-Haul truck with his belongings and the rest of Taylor's.

My recent trips driving to and from California had proved to be some of the worst of times, faced with miles of deserted road with nothing but time to think dark, lonely thoughts. The only thing different about this trip was that Joan and Taylor would be on the road next to me.

I'd expected Taylor to sleep the whole way because I'd seen her do that on trips to the boat, so it brought me some comfort to see her, snuggling like a little girl with her pink pillow against the passenger-side window, as Joan passed me occasionally on the highway to playfully point at my sleeping beauty. When she did, I shrugged, smiled, and gestured as if to say, "What else is new?"

I wondered what my life would be like with Taylor four hundred miles away in the City of Angels, which seemed like the perfect place for her after she'd been such a guardian angel to me.

Crossing into California, we made a quick stop in Indio for fuel and snacks. I could tell that Taylor was growing increasingly nervous because when I gave her a quick hug and kiss, she looked up at me with those beautiful blue eyes as if she was thinking, *How am I going to do this without you, Dad?*

I knew I needed to drive away before I started to cry, turned the car around, and told her she couldn't leave us just yet. I wanted to be strong for her, but I wasn't sure I would be able to once we got to her school the next day.

Once we arrived at the Westin Bonaventure Hotel in downtown Los Angeles, we decided to pick up some supplies for her apartment so we wouldn't be rushed the next day. We treated our stay like a minivacation to take our minds off the daunting task that loomed ahead like a four-hundred-pound gorilla that no one wanted to acknowledge.

Taylor had been assigned to a campus-leased apartment in a mixed-use business and residential area in West Los Angeles near the Wilshire business district, six miles from the campus and the Staples Center, with some gritty urban neighborhoods in between. We circled the neighborhood, identifying grocery stores, a dry cleaner, gas stations, and banks to help her feel comfortable in her new surroundings—and so we could feel comfortable about her being there.

Given that Taylor had never been away from us for an extended period of time, I wanted to make sure we found the safest route for her to take to school, which was at 9th and South Grand Avenue. I felt like a bodyguard, telling her how to keep watch while driving or walking, always looking for ways out of trouble. "Don't walk alone; go with a friend," I told her. "Know who's in front of you and who's behind you."

After feeling unsafe, unsure, and afraid for the past two years, I was an expert in this area. But I felt I needed to give her even more protection, so I stopped at a martial arts store to buy her a canister of pepper spray and told her how to use it. When I said I wanted her to carry it everywhere, she obediently put it in her purse.

"We'll have to try this on Anthony to make sure it's effective," I joked, which made us all laugh and eased the tension.

I could see by Taylor's face that she was lost in thought, and I assumed that she was feeling as I had for so long—overwhelmed by not knowing what to expect in a new environment.

"Are you sure I'm going to like it?" she asked.

I was determined to help her get through this. "At first it's difficult and overwhelming, but then this fear turns into curiosity, and then you start to embrace your new surroundings," I said. "You're going to become comfortable."

At the end of the day we headed back to the hotel suite, where we made up the pullout bed in the sofa for her to sleep on. I looked over at her, wrapped up in the pillows and blankets that she was so fond of, and wished that tomorrow would never come. Joan and I held each other tight in bed that night, just as upset as Taylor, wondering what our lives were going to be like without her.

Expecting a large crowd at the 10:00 A.M. school check-in, we got up at seven o'clock so we could get there early. Of course, I was ready first and waiting on Taylor, who always took forever to get dressed and made up, but this time I sat patiently in the living room while she prettied herself.

As I carried her suitcase and she toted her purse and other belongings down to the car, I felt we were delivering her to a future filled with excitement, new friends, and a new beginning. Set free of the daily drama of my recovery and her brother's addiction issues, she could finally concentrate on herself.

It was our job now to relieve her of those burdens, along with any residual guilt or sorrow. For whatever reason, I believed the new Scott could sense exactly what she was feeling, and today, that was emptiness. Even though she would be around lots of people, I suspected that she would feel alone for some time to come, just as I had.

. . .

After several hours of registering Taylor for school, we picked up her apartment keys and started moving her in. We were all pleased to find that her two-bedroom, two-bath unit on the twelfth and top floor was spacious, with giant windows facing southeast, overlooking the downtown skyscape. I was particularly happy to see a security guard out front with an intercom and all kinds of amenities in the enormous Park La Brea complex, which was like a small city.

Joan and I spent the day fixing up the kitchen, setting up the wireless Internet, putting up a ceiling fan, and doing some minor relocation of furniture. I enjoyed feeling needed and was more than willing to do anything Taylor asked—even take out the garbage. Soon her three roommates checked in, and the apartment began to get crowded with parents and students, so we wanted to let Taylor get acquainted with them and ready herself for the first night on her own.

We stayed in town for a couple more days in case she needed us, doing our own thing during the day and meeting up with Taylor and Anthony for dinner, sightseeing, and some "needed" shopping on Melrose.

Sunday, our last full day in town, came too quickly. Taylor and Joan did some last-minute errands, then mom gave daughter a cooking lesson—cutting up chicken breasts, sautéing them with spices for dinner, then freezing some for later use—while I stayed at the hotel to watch football and prepare myself for the final goodbye that evening. Only I couldn't keep my mind on the game, as I struggled to come up with something profound to tell Taylor.

When I got to her apartment around 4:00 P.M., Joan and Taylor were laughing with Anthony and her new roommates in the kitchen. In just two days I'd seen Taylor go from the little girl sleeping in the car to the grown woman I'd been watching develop over the past two years, who in her new surroundings wore a look of confidence and security.

Seeing that, I felt that now was the perfect time to go, leaving Taylor with people to support her through the rest of the evening in case she had a hard time. I asked Joan to come with me into Taylor's room, where she reluctantly agreed, then went to fetch our daughter.

"Taylor, come here a second," I said, bringing her into the bedroom with us. "We're going to go now, so you can spend time getting to know your roommates."

When her eyes welled up with tears, my heart sank.

I thought we were doing the right thing, but maybe it's too soon. Is she going to be okay?

Panicking a little, I wondered if we should have stayed longer, as we'd originally planned. All I wanted to do was ease her pain. "Come downstairs with us so we can say good-bye," I said.

The air was heavy in the elevator as the three of us headed down to the parking lot. It was a long, silent ride, and I could feel my heart breaking as it never had before. Even in the cool breeze outside, I felt hot and my stomach turned with every step that we walked toward the car when Joan suddenly burst out with, "I forgot my purse upstairs." Taylor laughed through her tears at the typical Joan behavior, which she'd been exhibiting all weekend.

While Joan went upstairs, Taylor and I held hands and walked toward the car, where I was finally ready to say my piece. "You're ready for this," I told her. "Mom and I are so proud of you and the woman that you've become and the woman that you are about to be. I'm so proud that we're able to provide you with the schooling of your dreams. You deserve this happiness."

Taylor cried the entire time I was talking, not saying a word, but she didn't have to. I knew what she feeling: this was what she wanted, and she would make the most of her education, but parting was still difficult. I reminded her that we were only an hour's flight away and that we could talk and text whenever she wanted to. We got out of the car when we saw Joan approaching, who, after giving Taylor a big hug, burst into tears. "I'm going to miss you so much. Be safe," Joan choked out, which made me break down crying too. I'd stayed strong for both of them for this long, but I couldn't do it anymore.

After Joan let go, I squeezed Taylor so tight I thought she was going to pass out, but I couldn't help myself.

Like my memory, she is going to be difficult to live without. How am I going to do this?

"Be safe and have fun," I said. "We'll always be here for you. Never forget that we love you and will drop everything if you need us."

Joan and I watched her walk away, wiping away tears. I tried to think of something to say to make Joan feel better, but I knew that was impossible, so I just sat there, wanting to go back upstairs and spend one last night with our daughter, which I knew wouldn't be good for any of us.

We drove away with a box of Kleenex at our side, and after talking several more times with Taylor that evening, we managed to calm her down by assuring her the pain of leaving home would get better with time.

After a restless few hours of tossing and turning, I sneaked into the living room around 4:00 A.M. and closed the bedroom door to watch TV, leaving Joan sound asleep. I got dressed, and around 4:45 I went downstairs to get some coffee in the lobby, where I watched the news, got a refill, and took it outside for a walk down Flower Street.

There were office buildings behind and in front of me, but the sidewalks and streets were empty and quiet. The sky was just starting to get light, and soon the slightest hint of orange started creeping up from behind the silhouette of the buildings. It was a little chilly in my short-sleeved shirt with the breeze of dawn breaking, but the coffee helped to warm me up. I turned a corner, and the road sloped down, opening up a panoramic view of the valley with a mountain range in the distance. As the expansive orange-yellow orb started to rise over the peaks, I felt its warmth take the edge off the chill.

I often went for a drive in the early mornings in Arizona to watch the sun rise over the mountains because I enjoyed the stillness of this time of day more than any other. But this sunrise was different, and it was special. It was the dawn of a new day, not only a new beginning for Taylor but also for Joan and me, who would finally be alone together in the house. For the first time since we'd begun saying good-bye to Taylor, I felt an inner peace. My life seemed fuller than it had been at any time since the accident, so much so that I no longer considered it "deleted." Now, if Grant could only pull himself out of his dark depths and stay in recovery, all would be right with the world.

As people began wandering out of the buildings and a few cars drove by, I decided to start walking back before my peaceful feeling was ruined. I strolled at an easy pace back to the hotel, where I found Joan tucked in bed, still asleep.

. . .

Later that morning, while we were in a business meeting about a potential speaking engagement, Joan got a call from Grant. She stepped out to take the call, and after the meeting she told me matter-of-factly that Grant had been hospitalized and released, with a prescription for antidepressants. He no longer wanted to use drugs or felt suicidal and was on his way to speak to his counselor.

We were on the highway driving back to Arizona when Grant called again, and after listening to Joan's side of their twenty-five-minute conversation, I could tell that she was cautiously optimistic, if not happy, about the situation, which she confirmed when she filled me in. Essentially, she said, this was the call we'd been hoping for, and that made me happy too.

"I want help," he told her, saying he was willing to go back into the rehab program he'd recently quit. "I want to be part of the family again. I'm going to die if I continue using drugs, and I don't want to die. I want to live, and I want to live without drugs."

Now that Grant had joined Taylor, Joan, and me in pursuing the lives each of us wanted to lead, it really had turned out to be a new day, not just for me but for my rejuvenated marriage and my family as a whole. And as we continued back to Gilbert that afternoon, all I could see ahead was a long stretch of open road, with endless opportunities for us all.

ACKNOWLEDGMENTS

This book was written with the help of many family and friends, most who have known me for many years and were truly there for me at a time when I did not know who I was.

I would like to give a special thanks to my parents, Louis and Alice Bolzan, who have provided me with so many memories of who I was when I was a young man and for instilling in me the values and integrity that have allowed me to carry on. To Joan's parents, Harvey and Fran Clack, for their endless support, prayers, and love. To our loving children, Grant and Taylor, your love and limitless caring has made me realize why I need to move forward being the father that I once was and how to become an even better father now. Thank you to Kevin, Jaime, Noah, Aden, and Luke Southard for our times together and sharing in the birth of Luke that allowed me to laugh and feel close to family again. To my cousin Brad Budner, who took the time to uncover and allowed me to relearn some of our family history and treasured memories.

To Mattie Cummins, who gave me the opportunity to believe I could not only move forward but I could be there to help others. To my teammates from Northern Illinois University, Brendan Dolan and Darren Stahulak, who have re-created so many of the new memories of my college football days and for giving me an insight into what it must have been like being a student athlete. To Phil and Linda Herra, who took time out to show us around NIU and provided countless stories to piece my life together. To Scott Kellar, Terry Clemans, and Vince Scott for their emails in time of need, that when I was struggling to figure out who I was as a man provided me stories of when I was a captain for our team. You will never know how much that meant to me. To Coach Bill Mallory, I would like to thank you for your personal calls and in the four years you coached me not only to become a champion and NFL player but teaching me how to be a better man.

To my dear friend Mark Hyman who has been there for me, who gave me so much emotional support and allowed me to share my darkest days and always gave me words of encouragement.

We would like to extend a heartfelt thank you for all of the above people who were such support for Joan as well and for that we are eternally grateful. Next, both of us would like to give huge thanks and hugs (from Joan) for the devoted support from our loving friends: To Joan's best friend Karen Peterson, who provided us with key brain injury research and allowed Joan to vent, cry, laugh, gave her nonstop help, and for that we are forever appreciative. To Randy, Johnna, and Justin Leach, who were there for us and gave us authentic emotional and spiritual support and many needed laughs. To Robyn Rieger for your touching inspiration and compassionate encouragement. To Diane Wallace for sharing her photography talents and caring friendship with us. To Dr. Rich and Kathy Silver, who compassionately welcomed us into their lives and provided medical support and so many good new memories. To Dr. Theresa Lanier, Sheryl Acevedo, and the staff at Arizona Vista Family Medicine, who provided a lifeline for us emotionally and medically with every step of this journey.

In advance, to Suzanne Wickham, whose dedicated promotion will allow us to share our story with the world. To Nancy Hancock, our amazing editor, who has guided us through this journey with such professionalism and kindness and skillfully took our words deeper so we could share it on a personal level with every reader. To Christina Bailly and the entire staff of HarperCollins, who embraced our story with open arms and utter support beyond our wildest dreams.

Finally, thank you to all the people who reached out with support, prayers, and kind words. Please forgive us if we *forgot* to mention you by name, but your encouragement and thoughtfulness will remain forever in our hearts.

A portion of the authors' earned royalties

from sales of this book will be donated to the

Brain Injury Association of Arizona and the Phoenix

Children's Hospital: Neuro-NICU Department.